LIST OF TITLES

Already published

A Biochemical Approach to Nutrition	R.A. Freedland, S. Briggs
Biochemical Genetics (second edition)	R.A. Woods
Biological Energy Conservation (second edition)	C.W. Jones
Biomechanics	R.McN. Alexander
Brain Biochemistry (second edition)	H.S. Bachelard
Cellular Degradative Processes	R.T. Dean
Cellular Development	D.R. Garrod
Cellular Recognition	M.F. Greaves
Control of Enzyme Activity (second edition)	P. Cohen
Cytogenetics of Man and other Animals	A. McDermott
Enzyme Kinetics (second edition)	P.C. Engel
Functions of Biological Membranes	M. Davies
Genetic Engineering: Cloning DNA	D. Glover
Hormone Action	A. Malkinson
Human Evolution	B.A. Wood
Human Genetics	J.H. Edwards
Immunochemistry	M.W. Steward
Insect Biochemistry	H.H. Rees
Isoenzymes	C.C. Rider, C.B. Taylor
Metabolic Regulation	R. Denton, C.I. Pogson
Metals in Biochemistry	P.M. Harrison, R. Hoare
Molecular Virology	T.H. Pennington, D.A. Ritchie
Motility of Living Cells	P. Cappuccinelli
Plant Cytogenetics	D.M. Moore
Polysaccharide Shapes	D.A. Rees
Population Genetics	L.M. Cook
Protein Biosynthesis	A.E. Smith
RNA Biosynthesis	R.H. Burdon
The Selectivity of Drugs	A. Albert
Transport Phenomena in Plants	D.A. Baker
Membrane Biochemistry	E. Sim
Muscle Contraction	C. Bagshaw
Glycoproteins	R.C. Hughes

Editors' foreword

The student of biological science in his final years as an undergraduate and his first years as a graduate is expected to gain some familiarity with current research at the frontiers of his discipline. New research work is published in a perplexing diversity of publications and is inevitably concerned with the minutiae of the subject. The sheer number of research journals and papers also causes confusion and difficulties of assimilation. Reviews articles usually presuppose a background knowledge of the field and are inevitably rather restricted in scope. There is thus a need for short but authoritative introductions to those areas of modern biological research which are either not dealt with in standard introductory textbooks or are not dealt with in sufficient detail to enable the student to go on from them to read scholarly reviews with profit. This series of books is designed to satisfy this need. The authors have been asked to produce a brief outline of their subject assuming that their readers will have read and remembered much of a standard introductory textbook of biology. This outline then sets out to provide by building on this basis, the conceptual framework within which modern research work is progressing and aims to give the reader an indication of the problems, both conceptual and practical, which must be overcome if progress is to be maintained. We hope that students will go on to read the more detailed reviews and articles to which reference is made with a greater insight and understanding of how they fit into the overall scheme of modern research effort and may thus be helped to choose where to make their own contribution to this effort. These books are guidebooks, not textbooks. Modern research pays scant regard for the academic divisions into which biological teaching and introductory textbooks must, to a certain extent, be divided. We have thus concentrated in this series on providing guides to those areas which fall between, or which involve, several different academic disciplines. It is here that the gap between the textbook and the research paper is widest and where the need for guidance is greatest. In so doing we hope to have extended or supplemented but not supplanted main texts, and to have given students assistance in seeing how modern biological research is progressing, while at the same time providing a foundation for self help in the achievement of successful examination results.;

General Editors:

**W.J. Brammar, Professor of Biochemistry,
University of Leicester, UK**

**M. Edidin, Professor of Biology
Johns Hopkins University, Baltimore, USA**

The Biochemistry of Membrane Transport

I.C. West

Lecturer, Department of Biochemistry
University of Newcastle upon Tyne

Chapman and Hall
London and New York

First published 1983
by Chapman and Hall Ltd
11 New Fetter Lane, London, EC4P 4EE
Published in the USA
by Chapman and Hall
733 Third Avenue, New York, NY 10017

© 1983 I.C. West

Printed in Great britain by
J. W. Arrowsmith Ltd., Bristol

ISBN 0 412 24190 0

This paperback edition is sold subject to the condition that it shall not, by way of trade or otherwise, be lent, resold, hired out, or otherwise circulated without the publisher's prior consent in any form of binding or cover other than that in which it is published and without a similar condition including this condition being imposed on the subsequent purchaser.

All rights reserved. No part of this book may be reprinted, or reproduced, or utilized in any form or by any electronic, mechanical or other means, now known or hereafter invented, including photocopying and recording, or in any information storage and retrieval system, without permission in writing from the publisher.

British Library Cataloguing in Publication Data

West, I. C.
   The biochemistry of membrane transport. —
   (Outline studies in biology)
   1. Biological transport   2. Cell membranes
   I. Title   II. Series
   574.87'5   QH601

   ISBN 0-412-24190-0

Library of Congress Cataloging in Publication Data

West, I. C. (Ian Charles)
   The biochemistry of membrane transport.
   (Outline studies in biology)
   Bibliography: p.
   Includes index.
   1. Biological transport. 2. Membranes (Biology).
I. Title. II. Series: Outline studies in biology
(Chapman and Hall) [DNLM: 1. Cell membrane . . Physiology.
2. Biological transport. QH 601 W517b]
QH509.W47 1983 574.87'5   83-7873
ISBN 0-412-24190-0

Contents

1 Introduction	**7**
1.1 The purpose of this book	7
1.2 The limiting membrane	7
1.3 Some definitions	10
1.4 Under what circumstances is active transport necessary or advantageous?	10
1.5 Common features between organisms	12
2 The kinetics of transport	**13**
2.1 The purpose of kinetic studies	13
2.2 Kinetics of simple diffusion	13
2.3 Kinetics of facilitated diffusion	17
2.4 Kinetics of coupled transport or secondary active transport	23
2.5 The kinetic effect of an electric field	24
3 Facilitated diffusion systems	**26**
3.1 Porins	26
3.2 Anion-exchange carrier of the human erythrocyte	29
3.3 The glucose carrier of the human erythrocyte	35
4 Secondary active transport	**38**
4.1 Lactose permease defined genetically	38
4.2 Lactose-proton symport defined physiologically	38
4.3 The lactose carrier protein	40
4.4 The mechanism of lactose-proton symport	43
5 Primary active transport systems	**47**
5.1 $(Na^+ + K^+)$-ATPase	47
5.2 Bacteriorhodopsin	57
6 Transport in mammalian metabolism: the control of transport and transport diseases	**62**
6.1 Gastric acid secretion	62
6.2 Diabetes mellitus and the normal function of insulin	65
6.3 Cholera	66
6.4 Vision	67
6.5 Control of ion-flow through nerve membranes	69
6.6 Hereditary defects of transport	72

Appendix	73
References	75
Index	79

1 Introduction

1.1 The purpose of this book

This short book is intended to outline the biochemist's view of biological transport. Though sharp distinctions between scientific disciplines are artificial, the biochemist's view is nevertheless different from those of the physiologist or the cell biologist. The biochemist is primarily concerned with discovering the molecular nature of the structures and the molecular details of the reactions through which these structures cycle or oscillate. His techniques involve isolation and purification, size and shape determinations, the identification of reactive groups, reaction stoichiometries and conformational changes, and the determination of the kinetics of identifiable chemical and conformational changes. The biochemist remains, of course, a biologist and should be aware of the purpose and the consequences of transport systems in the life of the organism. There have, however, in the last year or two been great advances in the biochemistry of this topic which for years could only be studied in a physiological manner, treating the transport mechanism as a black-box whose properties could only be inferred from its overall behaviour. It is these recent advances that justify the present short book, which is intended as a source and guide for third year undergraduates.

1.2 The limiting membrane

All cells are bounded by membranes. The early microscopists distinguished the membrane as the physical boundary, or skin, of the cell, but when the soluble constituents of the cell were identified it became clear that the cell membrane must represent a permeability barrier between the internal and the external aqueous milieux.

Accepting that the limiting membrane does present a permeability barrier to the movement of many solutes, one can proceed to ask which solutes, and why? In fact, it was an investigation of this question that gave Overton [1] the first clue as to the chemical nature of the membrane and led him to postulate a hydrophobic lipoidal barrier round the cell.

This is not the place for a description of membrane structure, as that is adequately covered in standard textbooks and monographs (e.g. [2]), but a few comments will be made about the permeability and transport properties of the structures that are currently assumed to represent the typical biological membrane. These comments are summarized in Fig. 1.1.

The concept of hydrophobicity has been thoroughly discussed by Tanford [3]. Hydrophobicity contributes not only the cohesive force that stabilizes the phospholipid bilayer and holds the integral proteins in place;

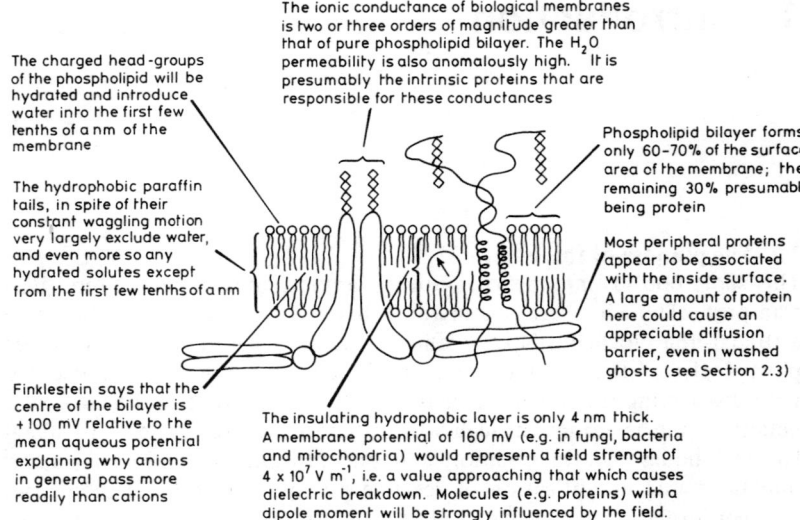

Fig. 1.1 Contemporary views concerning the properties of biological membranes.

it also provides the energy barrier (entropy barrier) that prevents H_2O from penetrating the fairly close-packed phospholipid tails of the bilayer. To remove one mole of water from the bulk aqueous medium and place it in the liquid parafin environment of the centre of the bilayer would require 22.5 kJ mol^{-1}. This value considerably exceeds the value of kT, the thermal energy of molecules (2.48 kJ mol^{-1} at ambient temperature), so H_2O is excluded from the centre of the bilayer. With H_2O excluded it follows that hydrophilic particles, e.g. ions and molecules that are strongly hydrogen-bonded to water (such as monosaccharides, disaccharides, etc.), will also be excluded. By contrast, there will be relatively little barrier to the passage of small hydrophobic molecules, such as CH_3OH, C_2H_5OH, CH_3COCH_3, as the membrane is exceedingly thin.

The membranes of many cells possess extra layers of a coarser and stronger nature than the fragile and vanishingly thin phospholipid bilayer, yet it is the latter that is the osmotic barrier round the cell and which can often sustain (at least over areas of a few hundreths nm^2) hydrostatic pressures of 30 atmospheres and ionic activity gradients equivalent to more than 10^7 V m^{-1}.

Although it is generally accepted that the existence of a limiting membrane, having restricted permeability to small hydrophilic molecules, is a prerequisite for life [4], not every scientist agrees with this view. There is a line of argument, vigorously pursued by Ling [5], Troshin [6] and others, which claims that the limiting membrane, while impermeable to proteins, is freely permeated by ions and small hydrophilic molecules, and that these latter only stay in the cell because they are specifically adsorbed by the immobile proteins, DNA, etc. If the conventional view sees the cell as soup in a polythene bag, this alternative view sees it as a sponge in a hair-net.

(See Kleinzeller and Kotyk [7] for a lively debate between the protagonists of the opposed theories.)

The experimental basis for the sponge-theory is slender. It stems from the belief that such important properties of the cell as permeability, active transport and excitability, should be properties of the 'vital' protoplasm and not of the indetectably thin and rather inert membrane. It has been pointed out that cells do not behave as perfect osmometers and it has been suggested that 'all the water in the protoplasm is somehow organized ('bound')' (Troshin [6]). It is also claimed that the mobility of certain ions is greatly reduced in cytoplasm compared with their mobility in dilute aqueous solution, so it is argued that the raised K^+/Na^+ ratio in cytoplasm compared with extracellular medium is due to selective binding of K^+, etc., by cytoplasmic constituents.

However, the sponge-theory cannot, or cannot easily and naturally, account for the following observations.

(a) Physical disruption of the membrane, e.g. with detergents or by osmotic shock, allows release of accumulated material.
(b) The cell, bounded by its limiting membrane, behaves as a perfect osmometer with respect to certain solutes.
(c) It was shown many years ago that galactosides, which can be accumulated to a high concentration in *Escherichia coli*, were osmotically active and were therefore not bound by adsorption [8].
(d) Enzymic activity inside a cell membrane can often be 'cryptic' or hidden in the absence of a means of ready ingress of the extracellular substrate. Catalysts of ingress have been found to be located in the cell membrane.

There are few contemporary biophysicists who do not accept that the membrane is an osmotic barrier. However, there are a number who believe that the conventional view greatly underestimates the extent to which H_2O is bound to, or at least partially affected by, the macromolecules of the cell. Thus, Brnjas-Kraljević claims that at the high *in vivo* concentrations of haemoglobin found in the red cell (7 mmolal), 1 g of haemoglobin affects the mobility of 0.5 g of H_2O. As the dry matter of cells can comprise more than 25% of the wet weight, it is clear that an appreciable fraction of the cell water is affected. Whittmann says that there is no need to invoke bound water to explain the odd osmotic behaviour of red cells, it is just that the osmotic coefficient (3.25 at 7 mmolal Hb) is higher than expected, but this says no more (nor less) than that haemoglobin reduces the activity of red cell water more than expected.

It is also emerging that certain types of solutes are selectively attracted into the hydration shell of macromolecules while other types of solute are excluded. The doubts of the sponge-theorists turn out to be well founded but greatly exaggerated. The overwhelming weight of evidence still justifies a rather simple soup-in-a-bag approach though refinements may become necessary to explain quantitative aspects such as diffusion constants, thermodynamic potentials and activity coefficients.

1.3 Some definitions

1.3.1 Simple versus facilitated diffusion
The story of transport across biological membranes is therefore about the departures from that overall picture of relative permeability towards hydrophobic substances, relative impermeability towards hydrophilic substances. Those molecules like O_2 and CO_2 that have to cross membranes at high rates, but which can do so unaided by virtue of being small and poorly solvated, no longer concern us, though the kinetics of this simple diffusion will be considered in Chapter 2. Mechanisms have evolved to facilitate the diffusion of specific hydrophilic molecules of which the diffusion would otherwise be too slow for the needs of the cell. We distinguish, therefore, between *simple diffusion*, where the pathway of diffusion is through the phospholipid bilayer, and *facilitated diffusion*, where diffusion is specifically accelerated by a specific combination between the solute and a component in the membrane, normally a protein.

1.3.2 Active versus passive transport
A diffusion process, whether catalysed or not, inevitably operates in one direction only, namely carrying solute from a place where its electrochemical potential is high to one where it is low. Not all transport systems in living organisms operate in this way. Many transport processes occur where there is a net movement of the transported solute from a low electrochemical potential to a higher potential. Such a transport is called *active* and we here distinguish between this active transport and the former *passive* transport (i.e. diffusion, whether facilitated or simple).

1.4 Under what circumstances is active Transport necessary or advantageous?
Active transport has posed some interesting problems for the theoretician, problems of an almost philosophical nature concerning the possibility of molecular machines operating in a thermodynamically reversible manner (i.e. with conservation of energy). Yet they seem to exist. There will be more discussion in a later chapter on the mechanisms of active transport. However, at this stage we might well consider the biological purpose (or purposes) of active transport.

1.4.1 Bilge-pumps
Any normal boat has to have some means of expelling unwanted water from the inside, which may have got there as spray or by slow seepage through gaps in the planking or riveting. Such devices we may call bilge-pumps. A cell bathed in a medium differing in composition from the cytoplasm will inevitably fill up with ions and other solutes from the outer medium, due to the imperfect diffusion barrier presented by the thin plasma membrane, and so active 'bilge-pumps' will be required. (The metaphor is Peter Mitchell's [9].) Thus, most animal cells and many parasitic and marine bacteria will be bathed in at least 140 mM Na^+. The

cytoplasm of these, as of most other living organisms, contains in the region of 10 nM Na^+ and the membrane potential would also favour Na^+ inflow. Consequently, here, as in most living organisms, Na^+ must be extruded by active transport. Calcium is also actively extruded from essentially all living organisms though the reason for its unwanted inflow is more to do with the negative membrane potential (interior negative with respect to the medium) and the two positive charges on the Ca^{2+} ion. A membrane potential of -60 mV would come into equilibrium with a 10-fold accumulation of Na^+ ion but a 10^2-fold gradient of Ca^{2+} ion. (See the Nernst equation.)

1.4.2 Energy storage or transduction
Bacteria, fungi and higher plants, as well as the bacterioid organelles (mitochondria and chloroplasts), actively transport H^+ ions outwards across their limiting membranes. The results are (i) a negative potential inside relative to the outside and (ii) a concentration gradient of H^+ ions. These effects sum to produce a strong tendency for H^+ ions to flow back inwards across the membrane. Such a flow can be harnessed to drive other processes. Thus, it can be used to synthesize ATP from ADP and P_i, as is found in aerobic bacteria, mitochondria and chloroplasts. However, in many organisms such as plants, fungi and anaerobic bacteria, ATP hydrolysis is used to establish the H^+ potential gradient in the first place and so the resynthesis of ATP is unlikely to be the objective. The transport of various other substances such as sugars, cations and anions, both into and out of the cell, can be driven by being coupled to this flow of H^+ ions. In animals the Na^+ potential gradient established by ATP hydrolysis is often used in a similar way.

1.4.3 Sequestration of activators
The contraction of actomyosin and the transmission of a nerve impulse across a synapse both depend upon the release of an activator (Ca^{2+}, acetylcholine, etc.) from a storage site to its site of action. The termination of contraction and transmission both depend on the rapid removal of the activator. In the case of muscle, the Ca^{2+} is sequestered by being pumped back into the lumen of the sarcoplasmic reticulum (from which it was originally released) by an ATP-driven pump. In the case of the neurotransmitters, the activators are actively transported first across the plasma-membrane of the synaptic terminal and then accumulated (again actively) in vesicles in the cytoplasm of the terminal (Fig. 1.2).

1.4.4 Scavenging food
Consider a bacterium in medium from which the nutrients are almost exhausted. The rate of assimilation of food determines the cell's survival. There are two procedures the cell could adopt: (i) Facilitate the passive entry of substrate so that it is present in the cytoplasm though at a concentration rather lower than in the medium. Subsequent entry into the metabolism of the cell could be made highly irreversible by the expenditure

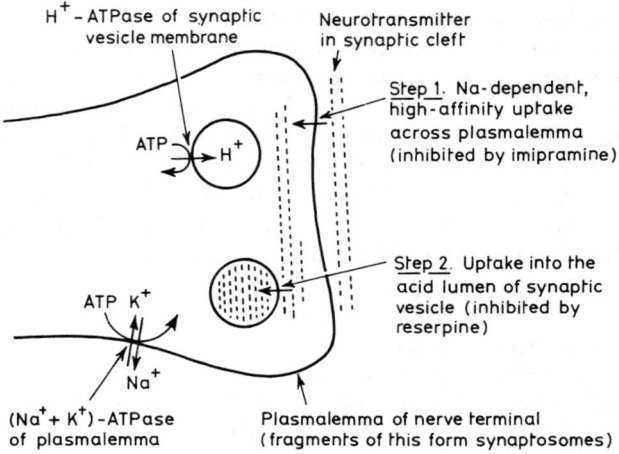

Fig. 1.2 A schematic representation of transport processes at a nerve synapse.

of energy but this step could be kinetically limited by the low concentration of substrate in the cell. (ii) Expend energy on active accumulation into the cytoplasm. Subsequent steps in metabolism could then occur with optimum substrate concentrations. In nearly every case the second strategy is adopted and it may be that it offers kinetic advantages. The only exception I know of is glycerol uptake in *E. coli*, which is via a pore [10].

1.4.5 Regulation of pH
The H^+ ion is actively transported out of most animal cells to maintain a cellular pH of around pH 7.4 [11]. Likewise, the acid lumen of the stomach and lysosome is maintained by active H^+ transport.

1.5 Common features between organisms
It may be asked, why isolate transport as a topic for consideration? Why not consider uptake in the gut in the context of nutrition, in the neurone as an aspect of excitability, in the mitochondrion in the context of energy transduction, and so on? In so far as there is an answer to that question, it is that transport processes in various organisms and in various tissues, cells or organelles, present a great many common features. An understanding of the mechanism in one case is enormously improved by a study of other transport processes. This point becomes very clear if one considers the H^+-coupled symport systems of bacteria, algae and fungi, and compares these in turn with the Na^+-coupled symport systems of animal cells; or if one considers the similarities between the acid-pump of the stomach and the $(Na^+ + K^+)$-pump of the plasmalemma; or the hormonal control of acid secretion in the stomach as compared with the hormonal control of glucose-uptake by fat-cells and muscle.

2 The kinetics of transport

2.1 The purpose of kinetic studies

Kinetic studies, interpreted broadly, include determining the overall turnover rate of an enzyme or a transport carrier, examining the change of rate with changing substrate or co-substrate concentration, with changing pH or temperature, and determining relative rates with different substrates and inhibitors. This type of investigation has been largely responsible for our present understanding of transport carriers. We can visualize a protein in the membrane, large enough to present a stereo-specific substrate–binding-site to the outer aqueous medium, capable either of 'floating' in the phospholipid bilayer and reorientating its binding-site or, more likely, of spanning the membrane and undergoing a conformational change so that the binding-site ceases to be accessible from the outer medium but becomes accessible from the inner medium.

Enormous progress is currently being made in determining the structure and the biochemical properties of transport carriers. Examples of this detailed knowledge of specific carriers will be presented in subsequent chapters. However, during the investigation of any transport process there comes a time to ask 'How does it work?' The answer may be something like: 'First this substrate binds and then that substrate,' or 'The slow conformational change between state 2 and state 1 is greatly accelerated by the binding of ATP, and thus a relatively high concentration of ATP is required'. These are the results obtained from kinetic experiments. In other words, the answer to the question 'How does it work?' turns out to be a kinetic model.

The subject of biochemical kinetics, while not being particularly difficult, does require a lot of concentration, and there comes a point for most students where further reading appears to be a waste of time. Certain of the fundamental concepts required here are, of course, common to enzyme kinetics, and these will not be enlarged upon. However, some attempt must be made to outline the type of experiment and the type of argument that underlies this kinetic approach.

It was remarked in Chapter 1 that facilitated diffusion is definable as an increase in the rate of transport over that expected for the process of simple diffusion of solutes across the phospholipid bilayer. It is therefore necessary to take a look first of all at the factors that govern the rate of simple diffusion.

2.2 Kinetics of simple diffusion

Rigorous treatments of diffusion will be found in some textbooks of physical chemistry or physical biochemistry. An excellent discussion of the subject from the point of view of membrane transport is given by Stein [12]. The treatment by Mitchell [13] is profound but difficult.

Consider the number of molecules per unit time (dn/dt) passing across a membrane of area A under a finite concentration gradient ($\Delta c/x$) where Δc is the difference in concentration across the membrane and x is the thickness (Fig. 2.1). Fick's empirical law can then be written:

$$\frac{dn}{dt} = -DA\frac{\Delta c}{x} \qquad (2.1)$$

The negative sign is because diffusion goes from high to low concentration. The diffusion coefficient, D, accounts for the different rates of diffusion of different compounds but, as these differential rates depend partly on the nature of the membrane, D must clearly also contain membrane-specific factors.

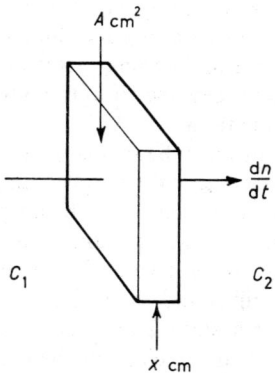

Fig. 2.1

By analogy with Ohm's Law, it will be seen that a flow is the product of a force times a conductance ($V = IR, I = V/R$). The force that causes diffusion, like the force that causes any other chemical process, is not strictly speaking a gradient of concentration (c), but a gradient of chemical potential (μ). These two quantities are related logarithmically ($c = e^{\mu/RT}$ or $\ln c = \mu/RT$ or $RT \ln c = \mu$).

If we write Fick's equation in its differential form:

$$\frac{dn}{dt} = -DA\frac{dc}{dx} \qquad (2.2)$$

and make the substitution $c = e^{\mu/RT}$

$$\frac{dn}{dt} = -DA\frac{de^{\mu/RT}}{dx}$$

therefore $$\frac{dn}{dt} = -DA\, e^{\mu/RT}\, \frac{d(\mu/RT)}{dx}$$

and $$\frac{dn}{dt} = -D\frac{A}{RT} c\, d\mu/dx \qquad (2.3)$$

This analysis of Fick's empirical law (following Mitchell [13]) shows that the rate of diffusion is proportional not only to the driving force ($d\mu/dx$) but also to the concentration of diffusate in the membrane (c), a fact that was not so obvious at first glance. We immediately see that this makes good sense and that we could have predicted it in a theoretical approach from the other end. If there are two molecules diffusing in the membrane we will get twice the flux expected from only one molecule.

Looking again at the analogy with Ohm's Law we see that the diffusion coefficient of Fick's Law (D) is a conductance and that we can replace it with its reciprocal, a resistance: in this case the resistance will be the frictional coefficient (f) for the interaction between the diffusate and the membrane. Thus (2.3) becomes:

$$\frac{dn}{dt} = -\frac{1}{f}\frac{A}{RT} c\, \frac{d\mu}{dx} \qquad (2.4)$$

This equation can give us some further insight into the factors determining the rate of simple diffusion. An amusing example occurs in the first three pages of Mitchell [13] where the diffusion of O_2 in muscle tissue is discussed. Given that myoglobin diffuses 100 times slower than O_2 (i.e. its frictional coefficient is 100 times bigger), how is it that myoglobin catalyses (increases the speed of) the diffusion of O_2 through muscle tissue? The answer to this is that the concentration of free O_2 is very low, the concentration of myoglobin is quite high (180 μM), its affinity for O_2 is high (K_D = 5.4 μM) so the concentration of oxy-myoglobin is high. Catalysis is achieved by raising c, the concentration of the diffusing species, and not by reducing the frictional coefficient (which in this case is considerably increased).

Let us now look at Collander's data following the treatment by Stein [12] (Fig. 2.2). The rates of diffusion of a range of chemical species across a biological membrane were measured. It was found that there was a fair correlation between oil/water partition coefficient and rate of diffusion. This is almost certainly due to the c term again; the rate of diffusion is higher for those compounds that partition into the lipid layer of the membrane. Independently of the oil/water partition coefficient, smaller molecules travelled faster. This is clearly a result of the frictional coefficient term. Stein showed that for diffusion in water, the frictional coefficient was proportional to $\sqrt{M_r}$ (the square root of the molecular weight) for small molecules and $\sqrt[3]{M_r}$ for proteins, but that for diffusion through polymers and biological membranes, the frictional coefficients depend much more steeply on relative molecular mass ($\propto M_r^{3.5}$)[14].

Stein also showed that the oil/water partition coefficient could itself be

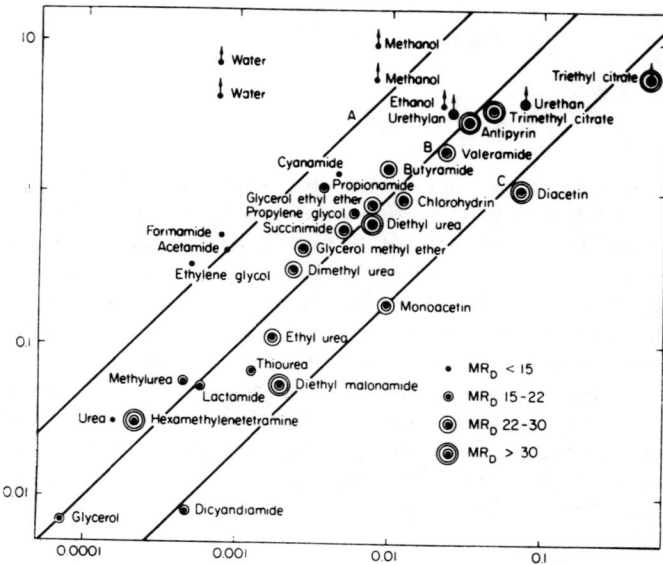

Fig. 2.2 The permeability of cells of *C. ceratophylla* to organic nonelectrolytes of different oil solubility and different molecular size. Ordinate: $PM^{1/2}$ (P, the permeability constant in cm h^{-1}; M, the molecular weight); abscissa: olive oil–water partition coefficients. MR_D is the molar refraction of the molecules depicted, a parameter proportional to the molecular volume (from [104] with permission).

roughly equated with another parameter, namely the number of hydrogen bonds formed between solute and solvent (water). The number of hydrogen bonds was ascribed to molecules according to Table 2.1.

Clearly, it is the energy of the hydrogen bonds that form between solvent and solute that determines partitioning into the phospholipid bilayer and therefore the rate of diffusion. At room temperature every

Table 2.1

Function	Chemical group in which it is present	Number of H-bonds assigned
H—O—H	water	4
—OH	alcohols, sugars, glycols, carboxylic acids	2
—NH$_2$	primary amines and amides	2
—NRH	secondary amines and amides	1
—C≡N	nitriles, dicyandiamide	1
—CO—	carboxylic acids, amides, aldehydes	1
—CO—	esters	½
—O—	ethers	0

kJ mol^{-1} of hydrogen bonding energy should lower c by a factor of 1.5 and therefore slow transport by the same factor ($e^{1000/RT} = 1.5$, at 20°C). This is found to be approximately the case.

2.3 Kinetics of facilitated diffusion

The rate of translocation of hydrophilic molecules across biological membranes could be enhanced in several ways, e.g.

(a) Increasing A by increasing the surface area of the membrane,
(b) Decreasing f by opening a pore through the membrane,
(c) Increasing $d\mu/dx$ by making the diffusion path shorter,
(d) Increasing c by formation of a specific complex between solute and a protein catalyst of translocation.

In this section we shall discuss only the *kinetics* of these mediated systems. In subsequent chapters other aspects of their biochemistry will be considered.

The first procedure is, of course, widely adopted but is, in a sense, trivial. Consider, for examples, the alveoli of the lung, the microvilli of the intestinal and renal brush-borders, chloroplast grana, and the mitochondrial cristae.

There are several examples of pores, but it is hard to find general kinetic features of these systems. Some pores are relatively unspecific and allow the ready passage of molecules smaller than a certain size. For the porins of the outer membranes of Gram-negative bacteria the cut-off point is at a relative molecular mass of around 700 (see Chapter 3) while the porins of the mitochondrial outer membrane show a cut-off around M_r 2000–6000. Some of the pores, while admitting free passage of a variety of small molecules, are rather selectively permeable to larger molecules, e.g. the maltose pore of *E. coli* outer membranes [29].

One theoretical treatment of the pore model is to calculate the equivalent radius of the pore from the permeability data of a variety of different sized substrates. However, this subject, well treated by Stein [12,15] and Schultz [16], is beyond the scope of this monograph.

A great deal of theoretical analysis has been devoted to the concept of a protein carrier molecule that forms a specific complex with the transported substrate on one side of the membrane and then undergoes a translocation step or a conformational change so that the binding-site becomes accessible from the other side of the membrane. This is called the *mobile carrier model*.

It is essential to emphasize at the outset that the following kinetic treatment cannot distinguish between a mobile carrier that floats across the membrane like a 'ferry-boat' and a larger transmembrane protein or oligomer of proteins enclosing a 'gated-pore' or substrate-specific channel. Both behave the same kinetically, both are examples of mobile carriers (Fig. 2.3).

The essential features of the mobile carrier model are:

(a) that there is a substrate-specific binding site;

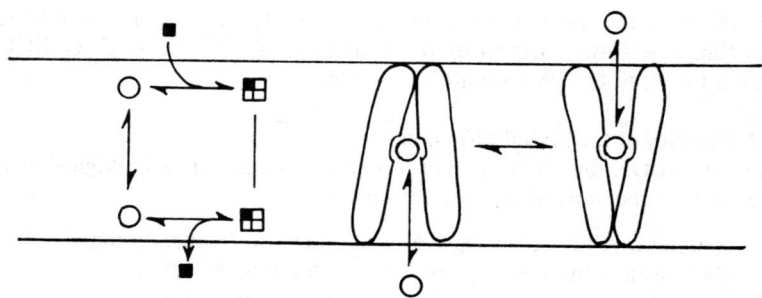

Fig. 2.3 Two extreme types of 'mobile carrier': the 'ferry-boat' (left) and the 'gated-pore'.

(b) that the binding site alternates between two states, in which it is exposed first on one side, then on the other side of the membrane;
(c) that the transition between these two states is a discrete kinetic event to which a rate constant can be assigned.

The great majority of facilitated diffusion systems in biology conform to this model and it will be fruitful to list its important properties.

(1) Facilitated diffusion by a mobile carrier system operates only so that solute flows from a higher to a lower electrochemical potential.
(2) Solute flows at a rate greater than that predicted from its size or hydrophilicity (discussed in the preceeding section).
(3) Rate of penetration does not follow Fick's Law except at very low concentrations. At higher concentrations saturation kinetics are observed (cf. enzyme kinetics).
(4) Competitive inhibition occurs between chemically and sterically similar substrates.
(5) Inhibition may be caused by other compounds, especially those reacting with or ligating reactive groups in proteins.
(6) It is often found that the rate of isotopic exchange at equilibrium is considerably greater than the rate of net flow. (This is called the exchange–diffusion phenomenon (Fig. 2.4).)
(7) It is often possible for the flow of substrate in one direction temporarily to drive the accumulation of an analogue (or isotope) in the opposite direction against the electro–chemical potential gradient of the analogue. (This is called the counter-transport phenomenon (Fig. 2.5).) (Does this contradict feature (1)?)

The last two properties, if they are demonstrable, constitute the strongest evidence in favour of the mobile-carrier model. They will be discussed again in more detail in a subsequent paragraph.

A high temperature coefficient (Q_{10}) is often cited as evidence of carrier-facilitated diffusion, but it is very weak evidence. Indeed, it is not really evidence at all, but a result of weak thinking. Any process with a high energy of activation will show a high temperature coefficient; for

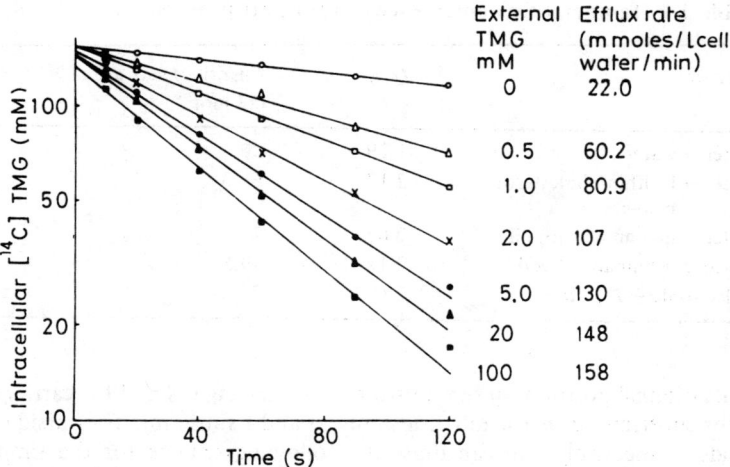

Fig. 2.4 Exchange–diffusion of galactosides in *E. coli*. The rate of loss of internal galactoside (^{14}C-TMG) is increased by increasing external galactoside (from [104]).

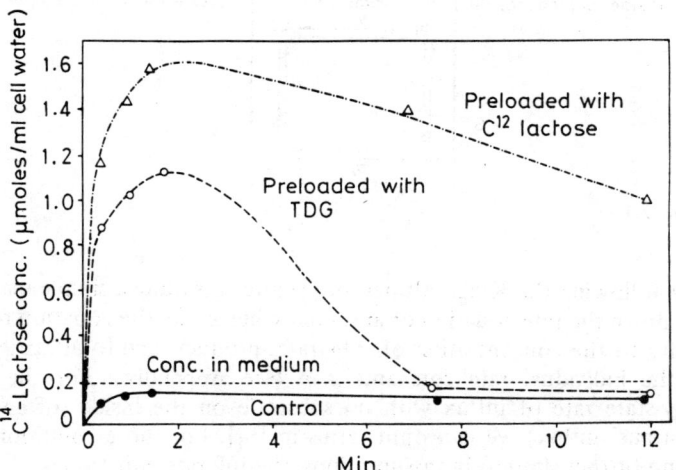

Fig. 2.5 Counter-transport of galactosides in *E. coli*. Lactose accumulates transiently into pre-loaded cell. As metabolism is inhibited the final equilibrium represents no accumulation (from [105]).

example, the opening of spaces between tightly packed phospholipid tails in the bilayer. The data in Table 2.2 shows that the simple diffusion of water into various media can have a higher energy of activation than the facilitated diffusion of glucose into red blood cells or the active transport of galactoside into *E. coli*.

The transport process can be modelled exactly as one models an enzyme reaction, with the provision of a single extra reversible step, namely the step returning empty carrier from the product side of the membrane back

19

Table 2.2 Temperature dependence of transport processes

Process	Q_{10}	Energy of activation (kJ mol^{-1})
Water → water	1.29	20
Water → lecithin/cholesterol bilayers	2.17	61.3
Water → human neutrophil	2.65	77
Glucose → human red cell	2.12	59.5
Galactoside → *E. coli*	2.60	75.6

to its original position on the substrate side. Letting C stand for carrier and S for substrate, a–h for rate constants, o and i subscripts for outside and inside, respectively, we can draw the following scheme for the simplest possible carrier-mediated uniport.

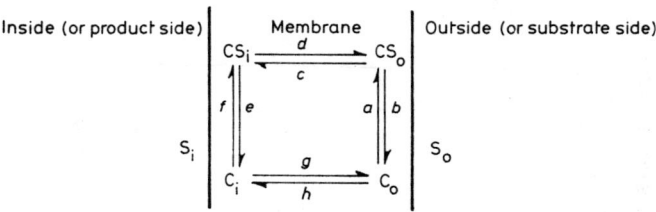

Scheme 2.1

By following the King–Altman procedure it is quite a simple matter to write down the rate equation of any such scheme; i.e. the equation relating velocity to the concentration of substrate, product, and total carrier (C_T) and the individual rate constants a–h (see Appendix). For the initial steady-state rate of influx with no substrate on the inside (often called 'zero-trans' influx) we can omit terms in [S_i]. For those conditions, but with no further simplifying assumptions, the full rate equation is:

$$v = \frac{C_T [S_o] aceg/\{a(ce + cg + dg + eg)\}}{[S_o] + (h+g)(bd + be + ce)/\{a(ce + cg + dg + eg)\}} \quad (2.5)$$

The model must be tested against reality, the properties of this equation matched against those of the transport processes under investigation. By comparing equation (2.5) with the well-known Michaelis–Menten equation:

$$v = \frac{V_m [S]}{[S] + K_m}$$

we see that a plot of initial rate of influx against external substrate concentration will have the usual hyperbolic form with a linear Lineweaver–Burk plot. It is also clear that the 'velocity' constant (V_m) and the 'concentration' constant (K_m) are the following clusters of intrinsic constants:

$$V_m = C_T \frac{ceg}{(ce + cg + dg + eg)} \tag{2.6}$$

$$K_m = \frac{(h+g)(bd + be + ce)}{a(ce + cg + dg + eg)} \tag{2.7}$$

Steady-state kinetic experiments yield only these two quantities, V_m and K_m, so it is clearly impossible to solve for nine unknowns. For simplicity it is often assumed that the binding reactions (a,b,e,f) are fast relative to the translocation reactions (c,d,g,h), which is equivalent to making the Michaelis–Menten assumption that $K_m = K_D$. It will be seen from Equation 2.7 that K_m will be dominated by the ratio b/a in that case. It is also often assumed, again for simplicity, that the carrier is symmetrical ($c = d = g = h$; $b = e$). However, these assumptions are usually unjustified and are often wrong.

As an example, let us take the glucose carrier of the human erythrocyte, which has been intensively studied and illustrates almost every aspect of the kinetics of the simple mobile carrier. Some values for K_m and V_m for a range of temperatures and substrates are given in Table 2.3. The change in V_m with temperature is to be expected (from Table 2.2), but the similar striking change of K_m with temperature probably indicates the invalidity of the rapid-binding assumption. It looks as though K_m is dominated not by the rate constants a and b, which you will note are missing from V_m, but by other rate constants that are found in the expression for V_m.

Table 2.3 Kinetic constants for the erythrocyte glucose carrier (from Stein [12]

Substrate	Temperature (°C)	K_m (mM)	V_m (mM in cell water min^{-1})
Glucose	37	10	198
Glucose	25	2.7	130
Glucose	5	0.5	5
Galactose	37	40	710
Mannose	37	13	710
Xylose	37	110	650
Ribose	37	2000	650
L-Sorbose	37	3100	124

K_m and V_m can be measured in several different ways. The zero-trans method has already been mentioned. Another way of measuring K_m and V_m involves allowing cells to equilibrate with certain concentrations of unlabelled sugar and then adding tracer to the outside. It is at first surprising to find that V_m measured in this way is twice that measured by the zero-trans procedure, until it is realized that in this 'equilibrium-exchange' experiment the steps g and h need not be used and indeed, at saturating substrate concentration, cannot be used. The above observation indicates that step d is faster than step g (assuming symmetry elsewhere).

A third technique, introduced by Sen and Widdas [18], has been called the 'infinite-cis' technique (for reasons that will become obvious). It was argued that if the cell is loaded with sugar to a concentration sufficient to saturate the carriers on the inside, and suspended in saline, sugar will flow out at V_m^{eff}. This V_m^{eff} will equal the zero-trans V_m for inflow if the carrier is symmetrical in this respect as it nearly is. If the concentration of the same sugar outside the cell is increased until *net* outflow is lowered to 50% of maximum, you could argue that influx is half V_m^{eff}, and therefore that the external sugar concentration is at 'K_m'. This, however, gives a different 'K_m' from the other two procedures and it will be noticed that the difference is that the step g of Scheme 2.1 does not occur at all in this procedure, even at intermediate external substrate concentrations.

Table 2.4 Kinetic parameters for glucose in the human erythrocyte at 20°C (data from [19,20].)

Procedure	Inflow		Outflow	
	K_m (mM)	V_m (mM in cell water min^{-1})	K_m (mM)	V_m (mM in cell water min^{-1})
Zero-trans	1.6–25	23	29	163
Equilibrium-exchange	28	300	28	300
Sen–Widdas (infinite-cis)	1.75	85	2.4	166

This concept, that c need not equal h, nor d equal g, is enormously important. The loaded carrier may reorientate faster than the unloaded carrier and often does. This is the explanation of the exchange-diffusion phenomenon. In this system the ratio between these two rates appears to be at least 2 and may, in fact, lie in the range 11–64 (Regen and Tarpley, [21]). In the case of the K^+ and H^+ specific ionophore Nigericin, the adenine nucleotide translocator of mitochondria and the band 3 anion-exchange porter of red blood cells (Chapter 3), the ratio approaches infinity; these porters do not catalyse uniport, only tightly coupled antiport.

The counter-transport phenomenon does not depend on the above asymmetry, but on a much simpler principle (though it is quite hard to prove formally [12]). The pre-loaded sugar competes on the inside of the membrane with the analogue or isotope. There is competitive inhibition of analogue efflux so a net inflow results. This persists as long as the pre-loaded sugar is at a higher concentration inside than outside, but as the pre-loaded sugar eventually equilibrates, so also does the net inflow of analogue reverse until it finally reaches equilibrium. It is usually possible, by judicious choice of analogues and conditions, to achieve a striking transient accumulation of isotope, which has made this a favourite way of confirming the existence of a 'mobile' transport carrier (Fig. 2.5).

In addition to the types of asymmetries that we have just been considering there is another effect that can produce kinetic anomalies and that it is important to understand. This is the result of unstirred layers of solvent adjacent to the inside or even the outside face of the membrane. The usual manner of investigating this question of unstirred layers is to compare the kinetics of inflow of a rapidly transported substrate and a slowly transported substrate. If carrier transport is very rapid (e.g. glucose into the human erythrocyte) a depletion of sugar can occur near the surface of the membrane, especially at low substrate concentrations. This does not affect a slowly transported substrate like sorbose. It turns out that there is a detectable diffusion barrier on the inside of the erythrocyte (perhaps due to the cytoskeleton or the high viscosity of the concentrated haemoglobin solution).

2.4 Kinetics of coupled transport or secondary active transport

Coupled transport is analogous with an enzyme reaction involving two substrates and two products. It is a conceptually simple matter to derive the rate equation for coupled transport by the same procedure as was used to derive Equation 2.5, but it is technically rather daunting because of the great number of terms involved (64 in the numerator and 512 in the denominator). Scheme 2.2 shows the model from which it is a simple matter to write down the King–Altman patterns. In this scheme the co-ion is designated A; it could be H^+ or Na^+.

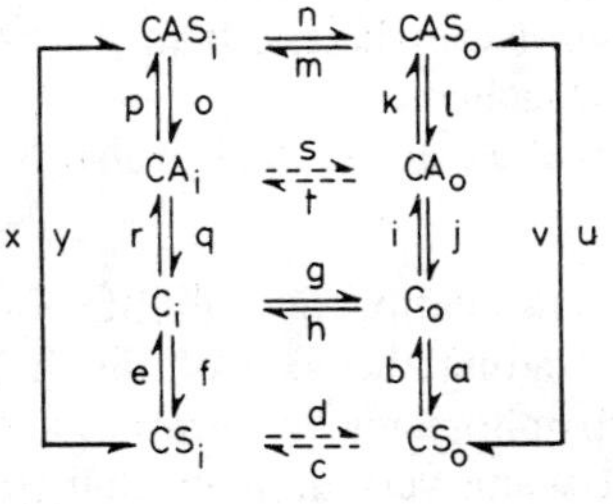

Scheme 2.2

Starting with C_o there are two possible routes for the formation of the ternary complex CAS_o, one via CA_o and one via CS_o. An 'ordered' mechanism would be one where only one of these routes was possible. A 'random' mechanism would exist if both were possible. The process described by rate constants c,d,s and t must be small or zero, otherwise slip would occur; the flow of S would not be tightly coupled to the flow of A, and vice versa. With a random mechanism it would be possible to say that the rate constants c,d,s and t must themselves be small, but with an ordered mechanism it could be that, for example, CS does not translocate because it does not form, rather than because c and d are small.

To distinguish between random and ordered mechanisms the standard methods of two-substrate enzyme kinetics can be applied. These are:

(a) isotope exchange at equilibrium,
(b) product inhibition studies,
(c) dead-end inhibitor studies,
(d) competitive inhibitor studies.

Page and West [22] conclude that the lactose permease of *E. coli* (see Chapter 4) with lactose as substrate presents an example of 'preferred' order, with proton binding first except at very high galactoside concentrations, when CS will form and slip will occur.

Steady-state kinetic studies cannot provide values for individual rate constants such as those in Schemes 2.1 and 2.2 (as previously remarked). To some extent these can be guessed at and the values put into the rate equation and tested for consistency with the data, but a large degree of uncertainty is likely to remain. The correct approach will eventually be to use the methods of single-turnover kinetics to examine individual rate constants, though this has not yet been successfully done for a transport process. Considerable data are now available on the single-turnover kinetics of $(Na^+ + K^+)$-ATPase as an enzyme, but not as an ion pump (See Chapter 5.).

2.5 The kinetic effect of an electric field

The thermodynamic effects of a membrane potential on the free-energy of ion movements are clear, and it is also clear that these must be the result of kinetic effects. However, these kinetic effects have proved quite difficult to understand. This subject rapidly grows too complex for an introductory text like the present one. Nevertheless, it might be valuable to show how this problem can be approached.

The kinetic effects of the electric field probably lie between the following two extremes.

(a) An effect on the translocation probabilities of charged carrier species (e.g. CA^+S or C^-) along the axis of the field. At its simplest this would be like electrophoresing the protein across the membrane. The drawback is that it seems very unlikely that there is a large movement of protein in that axis (Fig. 2.3). On the other hand Keith Wright has

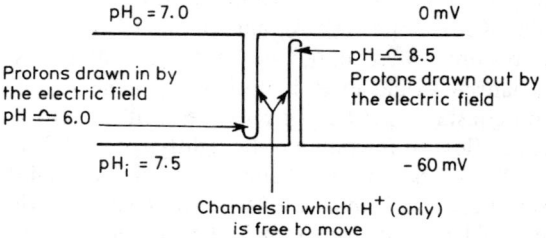

Fig. 2.6 Proton (and hydroxyl) wells.

recently pointed out that membrane proteins will, in general, have an electric dipole moment and that any conformational change could, in principle, change that dipole moment. The electric field will therefore influence conformational changes even though there may be no gross movement in the plane of the field.

(b) An effect on the concentration of charged substrates able to diffuse in the axis of the field. This latter effect of the field was first discussed (in terms of the proton) by Peter Mitchell [23], who pointed out that in an electric field a proton-specific infolding of the membrane would have a raised proton concentration at the bottom. He called this a 'proton-well'. A similar channel from the inside of the membrane would have a H^+-deficit at the bottom (Fig. 2.6).

3 Facilitated diffusion systems

3.1 Porins

The cell envelope of Gram-negative bacteria (and therefore of *E. coli*) is a complex structure comprising two distinct permeability barriers. Starting from the inside, one would first encounter the 'cytoplasmic membrane' (or 'plasma-membrane'), which contains the cytochromes, the dehydrogenases, oxidases, the ATPase and across which Mitchell's proton motive force is developed. This membrane is highly impermeable, even to that smallest of ions, the proton. The next structural feature is a mesh-work of peptidoglycan, the target of penicillin and lysozyme action. Outside that is a second phospholipid bilayer imbedded in which are some intrinsic proteins (including a lipoprotein) and lipopolysaccharide. The outermost layer of the cell is the polysaccharide chains of the lipopolysaccharide. This entire envelope is well illustrated in Fig. 3.1, from a review by Di Rienzo *et al.* [24].

The phospholipid bilayer of the outer membrane, if complete, would present the same sort of barrier to diffusion presented by the plasma membrane. The solution adopted by the Gram-negative bacteria is, however,

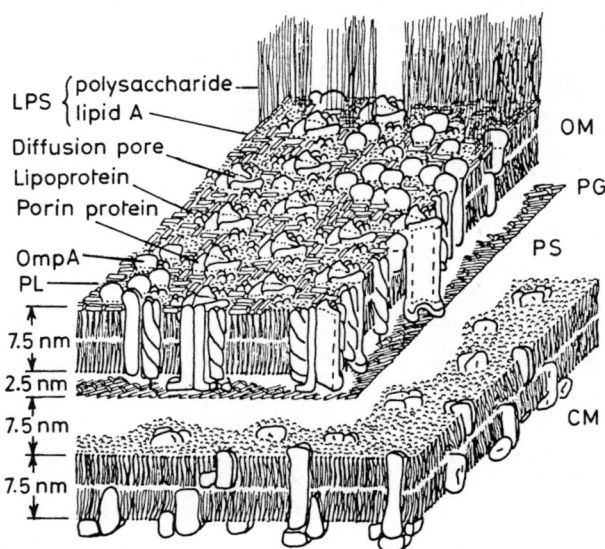

Fig. 3.1 A schematic representation illustrating the possible molecular architecture of the *E. coli* cell envelope. Abbreviations used are: PL, phospholipid; OM, outer membrane: PG, peptidoglycan; PS, periplasmic space; CM, cytoplasmic membrane. Polysaccharide chains in only some of the LPS molecules are shown (from [24] with permission).

quite different. It appears that there are pores through the outer membrane capable of passing hydrophilic molecules of up to 700–900 molecular weight. The protein monomers that form these pores (of which there are several types) are called porins.

3.1.1 Matrix protein porins

If the sonicated cell envelope of *E. coli* is spun in a sucrose density gradient, the outer membrane (density 1.22 g ml^{-1}) can be separated from the less dense cytoplasmic membrane (density 1.15 g ml^{-1}). Electrophoresis in the sodium dodecylsulphate polyacrylamide gel system of Laemmli reveals the following darkly stained bands in the region 40 000–26 000 M_r [25].

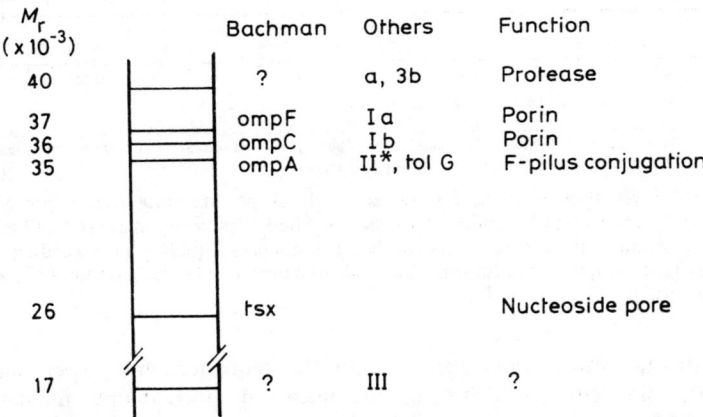

Fig. 3.2 Nomenclature of the major proteins of the outer membrane of *E. coli*.

The nomenclature has, until recently, been very confused and in any case it is not an important aspect of this topic. The mobilities of these bands differ in different electrophoretic systems. For example, ompA runs as 27 000 M_r if the sample is not boiled before applying to the gel. Proteins ompF, ompC and ompA are bound strongly, but not covalently, to the peptidoglycan.

The following lines of evidence indicate that proteins ompF and ompC, the so-called 'matrix proteins', form pores through the outer membrane, whence of course the name 'porin'.

(1) Nakae [26] showed that the incorporation of 'matrix protein' into artificial phospholipid vesicles greatly enhanced the permeability of these vesicles to sucrose. The reconstituted pores excluded oligosaccharides of greater than 900 M_r, as does the intact outer membrane.
(2) Some *E. coli* mutants, selected for resistance to Cu^{2+}, are found to lack the matrix protein from their outer membranes.
(3) Von Mayenburg's pleiotropic transport mutants show deficient transport at low concentration of a great variety of sugars, amino acids, bases and anions, and they also are found to lack the matrix proteins.

27

(4) Certain enzymes, such as 5' nucleotidase, are known to be located in the periplasmic space between the inner and the outer membranes. Mutant strains have been isolated where these enzymes are 'cryptic', and a M_r 41 000 protein was found to be missing from the outer membrane.

The closely related protein I from the strain *E. coli* B/r has been sequenced [27]. The molecular weight is 37 205. Charged residues are surprisingly evenly spaced along the molecule. There are no long stretches of hydrophobic amino acids, though many short stretches are to some extent clustered in the centre of the molecule (Fig. 3.3).

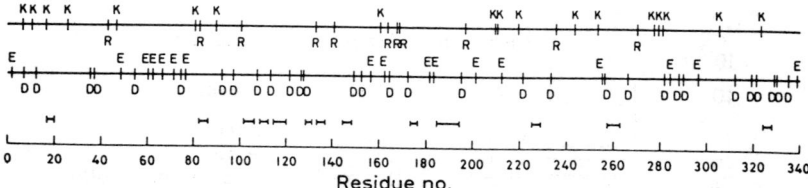

Fig. 3.3 Distribution of charged residues and hydrophobic sequences in protein I. Top line: lysine (K) and arginine (R) residues. Middle line: glutamate (E) and aspartate (D) residues. Bottom line: hydrophobic sequences equal to or exceeding four residues (the nonpolar residues alanine and glycine are included) (from [27] with permission).

Both theoretical calculations, from the sequence, and experimental studies, using circular dichroism and infra red spectroscopy, indicate a surprisingly low α-helix content (0–20%) and a surprisingly high β-sheet content (> 40%). This is surprising for a membrane protein, as those few about which we have any information appear to have a high α-helix content.

Electron microscope studies of matrix proteins mixed with lipopolysaccharide showed a hexagonal lattice structure. Cross-linking studies with the bifunctional amino group reagent dimethylsuberimidate ($H_3COC(NH_2^+)$-$(CH_2)_6C(HN_2^+)OCH_3$) revealed monomers, dimers, and trimers as well as faint bands of higher molecular weight. It is widely believed that each pore is constructed of three porin monomers, though it could be that the stable trimer has three pores. There seems to be no doubt that the monomer is without function.

The distribution of hydrophobic areas on the porin monomer makes more sense in the context of a trimeric pore, for specific folding of the peptide chain could result in a transmembrane cylinder having an hydrophobic sector and an hydrophilic sector. The trimer could then enclose a hydrophilic, water-filled pore that did not exist in the monomer (Fig. 3.4). This is rather speculative.

After adding very low concentrations of the matrix proteins into the aqueous medium bathing a black lipid membrane the conductance of the phospholipid bilayer begins to increase, but it does not increase steadily; it increases in discrete steps. Each conductance increment is interpreted as

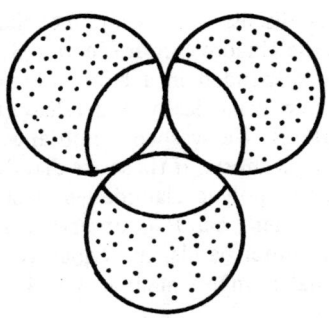

Fig. 3.4 Possible way in which the porin trimer could constitute one pore.

representing the conductance of a single pore. In experiments by Benz et al. [28] with an applied voltage of 20 mV and an aqueous solution of 1 M KCl, the typical increment was in the region of 1.3 nsiemens (1.3 × $10^{-9} \, \Omega^{-1}$). Benz et al. have calculated (somewhat naively) from this that the diameter of the aqueous pore is 0.9 nm (assuming that the length of the pore is 7.0 nm, and much else).

3.1.2 The lam B porin

This porin has recently been reviewed by Boos and colleagues [29]. Certain mutants of E. coli are resistant to λ phage because they lack the λ-receptor protein (i.e. they are lam B^-). These were found to show deficient growth on low concentrations of maltose. Further study showed that the λ-receptor, or lam B protein, has a molecular weight of 50 000. In the absence of matrix proteins, and when the cells are grown on maltose, it can be present at 10^5 copies per cell, as trimers, in the outer membrane. In the presence of the rather abundant matrix proteins it is possible that over-crowding prevents such a large number getting into the membrane. This porin, like the other proteins mentioned above, is non-covalently attached to the peptidoglycan network. It is responsible for the permeability of the outer membrane to maltodextrins of four to seven glucose units. Using the black-lipid-membrane technique described in the previous section, the pore is found to be of similar diameter to the matrix protein pores. However, this pore differs from the matrix protein pores in that, while it passes many smaller molecules and ions, it has considerable specificity for larger α1-4-linked glucose polymers. It binds these maltodextrins (and the even larger amylopectins) in vitro. This specificity towards maltodextrins is further enhanced by the close association in vivo between the lam B protein and the maltose-binding-protein found in the periplasm [29].

3.2 Anion-exchange carrier of the human erythrocyte

The human erythrocyte, available in large quantity in the form of outdated blood from blood-banks, has been an important cell for the investigation of membrane and transport phenomena. It provides us with two further examples of facilitated diffusion; both 'mobile carriers' rather than pores.

The first of these to be discussed is the anion-exchange carrier, or 'band 3' protein, the second is the glucose carrier (Section 3.3).

When red cells are suspended in a large volume of 5 mM phosphate buffer (pH 8) they burst osmotically and release their haemoglobin and other cytoplasmic proteins. The washed membranes are white, or very pale pink, and are referred to as ghosts. If these are dissolved in sodium dodecylsulphate and subjected to polyacrylamide gel electrophoresis, a relatively simple pattern of prominent bands is observed. These bands have been numbered 1–8. Some fainter bands, subsequently found to be important, have been given decimal numbers such as 4.1, 4.2, 4.5, etc. (see Section 3.3).

Band 3 has an apparent molecular weight of 95 000 and comprises 25% of the total membrane protein. However, band 3 is clearly composite. Purified $(Na^+ + K^+)$-ATPase is found to have a major subunit with a molecular weight of 95 000 (Chapter 5); but in the red blood cell Na^+-pumping is not a major activity, and there are only several hundred copies of this peptide per cell. Glycophorin, a heavily glycosylated, integral, membrane protein has an apparent monomer molecular weight of 40 000 to 50 000. It nevertheless forms such stable dimers that a considerable fraction of the cell's glycophorin is found in the 95 000 region. However, the greater part of the Coomassie Blue staining material in band 3 is a homogenous protein which has now been isolated, purified and characterized. It is this peptide that is referred to as the band 3 protein (for reviews see [30,31]).

3.2.1 Transport properties of band 3 protein

Band 3 protein has been identified as the site of anion transport in the following ways.

(1) Membrane ghosts can be depleted of extrinsic proteins by washing with NaOH. These depleted membranes retain band 3 and also retain anion exchange. Admittedly, small amounts of other proteins remain as contaminants after this treatment (see Section 3.3).
(2) The photo-affinity reagent N-(4-azido-2-nitrophenyl)-2-aminoethanesulphonate, nicknamed NAP-taurine (Fig. 3.5), is a substrate for the anion-exchange carrier and a competitive inhibitor of anion exchange. In the dark it slowly penetrates the cells, but in the light the unstable aryl azide group loses N_2 and the reactive nitrene reacts covalently with the first carbon atom it encounters. Subsequent analysis shows it mainly bound to band 3 protein.
(3) The rapid exchange of anions (Cl^-, HCO_3^-, $H_2PO_4^-$, SO_4^{2-}, etc.) across the red cell is strongly and very specifically inhibited by stilbene derivatives bearing charged sulphonate groups, e.g. 4,4'-diisothiocyano-2,2'-stilbenedisulphonate (DIDS; see Fig. 3.5). One of the sulphonate groups probably fits into the anion binding site, but these charged groups also serve to prevent the inhibitor from penetrating the membrane by simple diffusion. In the case of DIDS the isothiocyano group reacts covalently with the ϵ amino group of a lysine side chain, situated

(a) 4,4′-diisothiocyanostilbene-2,2′-disulphonate (DIDS)

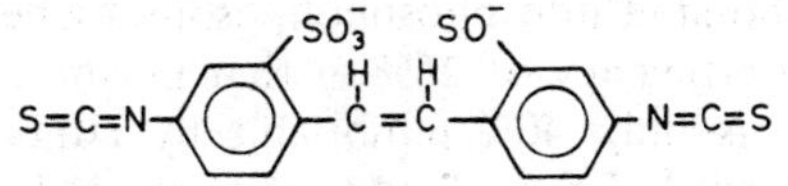

$$S=C=N-\underset{SO_3^-}{\bigcirc}-\underset{H}{\overset{H}{C}}=\underset{H}{\overset{H}{C}}-\underset{SO_3^-}{\bigcirc}-N=C=S$$

(b) 4,4′-diisothiocyanoditritiostilbene-2,2′-disulphonate ($^3H_2$ DIDS)

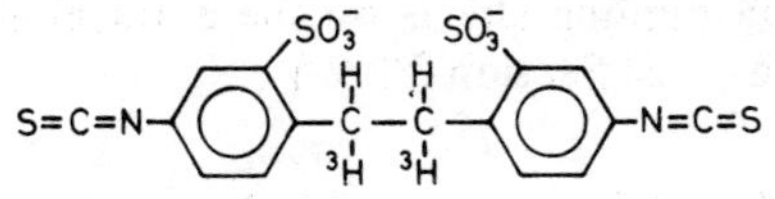

$$S=C=N-\underset{SO_3^-}{\bigcirc}-\underset{^3H}{\overset{H}{C}}-\underset{^3H}{\overset{H}{C}}-\underset{SO_3^-}{\bigcirc}-N=C=S$$

(c) 4,4′-diacetamidostilbene-2,2′-disulphonate (DAS)

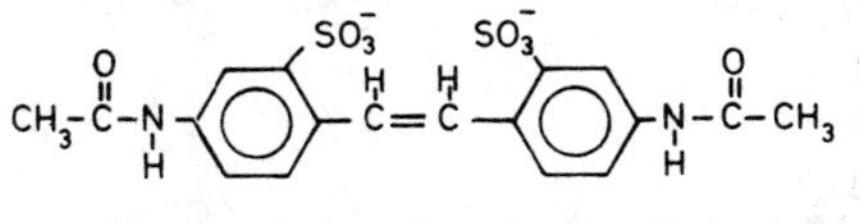

$$CH_3-\overset{O}{\overset{\|}{C}}-\underset{H}{N}-\underset{SO_3^-}{\bigcirc}-\underset{H}{\overset{H}{C}}=\underset{H}{\overset{H}{C}}-\underset{SO_3^-}{\bigcirc}-\underset{H}{N}-\overset{O}{\overset{\|}{C}}-CH_3$$

(d) N-(4-azido-2-nitrophenyl)-2-aminoethylsulphonate (NAP-taurine)

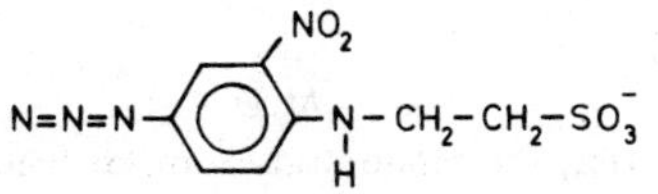

$$N=N=N-\underset{}{\bigcirc}\overset{NO_2}{\underset{}{}}-\underset{H}{N}-CH_2-CH_2-SO_3^-$$

Fig. 3.5 Some competitive inhibitors of anion transport in the red blood cell. (a) and (b) bind rapidly and reversibly but also react covalently. (c) binds but does not react covalently. (d) binds and when illuminated loses N_2 and reacts covalently. It slowly crosses the membrane and can react from the cytoplasmic surface, whereas (a) to (c) cannot cross the membrane and cannot react from the cytoplasmic surface even with inside-out vesicles.

near the binding site. Radiolabelled DIDS causes 99% inhibition of anion exchange when 1.2×10^6 molecules of DIDS have reacted per cell (Fig. 3.6). Independent estimates find an almost identical number of band 3 polypeptides per cell.

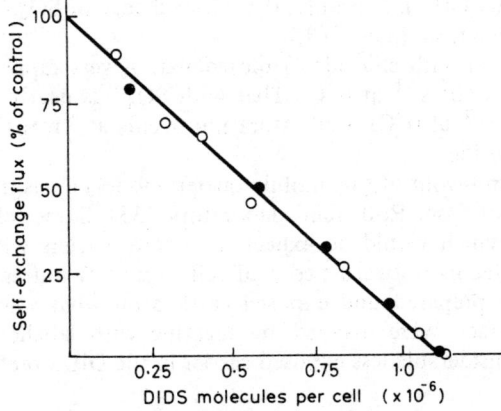

Fig. 3.6 Progressive inhibition of chloride (○) and bicarbonate (●) self-exchange in ghosts on reacting with DIDS (from [32] with permission).

(4) Band 3 protein can now be purified free of bands 4.2, 4.5 and glycophorin, and reincorporated into phospholiposomes where it catalyses anion exchange. An efficiency of 25% of that in whole cells has been achieved. Exchange is only 40% inhibitable by DIDS when that is added externally, but if DIDS is added to the protein before reincorporation into the liposome, 100% inhibition is obtained (Fig. 3.7). (Note that the DIDS binding site is on the extracellular face of the asymmetrical protein — see Section 3.2.2.)

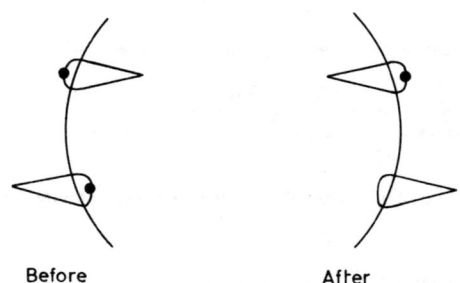

Before   After

Fig. 3.7 DIDS added before, or after, the random insertion of band 3 protein into liposomes.

Anion exchange is electroneutral. That must mean that there is a rather tight coupling between the inflow and the outflow of anions; i.e. the carrier is a strict antiporter (see Section 2.3). The simplest explanation is that the carrier operates as a 'mobile carrier' but the reorientation of the unloaded carrier does not occur (see Chapter 2). There is a small electrical conductance that is abolished by DIDS, so a slight 'slip' reaction may occur, a slight uncoupling between the inward and the outward fluxes; but it is very small compared with the rate of exchange.

As the carrier can exchange Cl^- for SO_4^{2-} but still be electroneutral, it would seem that SO_4^{2-} binding must be accompanied by H^+ binding (or Cl^- binding by OH^-). Certainly, the self-exchange of SO_4^{2-} has a different pH dependence from that of Cl^-.

The turnover with chloride or bicarbonate is very rapid (3.6×10^4 s^{-1} at 38°C, 2.7×10^2 s^{-1} at 0°C). That with SO_4^{2-} is nearly a thousandfold slower (0.34 s^{-1} at 0°C). Half-saturation occurs at 3 mM Cl^- outside and 65 mM Cl^- inside.

Evidence in favour of the 'mobile carrier' model comes from the following experiment from Rothstein's laboratory [33]. Some cells were treated with DIDS, which would be expected to trap carriers with their anion-binding site facing outwards; control cells were left untreated. Inside-out vesicles were prepared and exposed carriers on what was originally the cytoplasmic face were assayed by reacting with labelled NAP-taurine. There was considerable less exposed carrier in the DIDS-pretreated case.

3.2.2 Protein chemistry of band 3 protein

There is little doubt that the band 3 protein exists in the membrane as a stable dimer. Deliberate oxidation of −SH groups *in situ* in the membrane, with Cu^{2+}−o-phenanthroline, forms a dimer covalently linked by interchain S−S bridges. However, it is not necessarily the case that the functional unit is a dimer. Thus, each monomer binds the sulphonate derivatives, and cells with 80% of their monomers blocked with DIDS show 20% of the expected transport activity (Fig. 3.6). The monomers thus appear to act independently.

The complete amino acid sequence and the folding pattern of this protein is not yet known, but considerable information has been obtained by a number of now classical techniques (Table 3.1). Two features specific to this protein are the DIDS binding site already mentioned, and some attached carbohydrate residues which provide binding sites for several lectins including concanavalin A.

Table 3.1 Reagents for the investigation of integral membrane proteins

Reagent	Information obtained
Lactoperoxidase + $^{125}I_2$	Exposed Tyr iodinated, but the macromolecular reagent cannot penetrate the membrane.
Diazobenzenesulphonate	Reacts covalently with His, Cys, the anion etc, but the anion penetrates the membrane very slowly.
Trypsin, chymotrypsin, pepsin	Cleaves at specific amino acid residues but cannot penetrate the membrane.
Thermolysin	Cleaves at exposed peptide bonds rather unspecifically, but cannot penetrate the membrane.
Galactose oxidase followed by NaB^3H_4	Inserts ^3H into exposed galactosyl residues. Cannot penetrate membrane.
Cyanogenbromide	Cleaves at the carboxy side of Met.
2-Nitro-5-thiocyanobenzoic acid	Cleaves at amino side of Cys.
Hydroxylamine	Cleaves between Asn and Gly.
N-Bromosuccinimide	Cleaves at carboxy side of Trp and Tyr.

The sort of picture obtained by these studied is shown in Fig. 3.8. Important features to be noted are as follows: the N-terminus is on the cytoplasmic face and is blocked (acetylated); gentle trypsin treatment from the cytoplasmic face removes the N-terminal 41 000 M_r peptide, which is hydrophilic and which bears the Cys responsible for the S−S bridge; this treatment leaves a shorter, but still functional anion carrier in the membrane; gentle chymotrypsin treatment from the outside cleaves off the C-terminus as a 35 000 M_r peptide containing the lectin binding site(s). After both these treatments there remains a 15 000 to 19 000 M_r peptide spanning the membrane and bearing the DIDS binding site. Hydroxylamine cleaves a 10 000 M_r peptide off the C-terminus of this membrane-bound peptide which contains the DIDS site and the majority

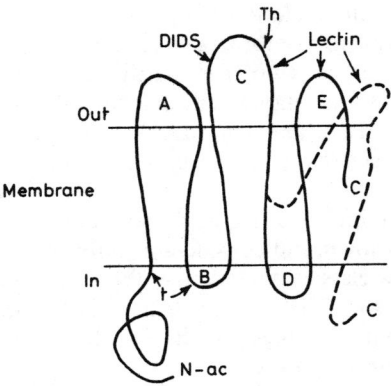

Fig. 3.8 Disposition of band 3 peptide across the ghost membrane (from [34] with permission).

of its externally iodinatable tyrosines. From the distribution of ^{125}I in the 14 small peptides obtained by thermolysin treatment, Tanner and colleagues [34] could infer that the band 3 protein traverses the membrane at least four, and probably six, times (Fig. 3.8).

The N-terminal hydrophilic segment, not required for anion transport, appears to anchor the band 3 protein to the cytoskeleton (and binds it to haemoglobin, and several glycolytic enzymes as well). Removal of this cytoplasmic portion of the molecule greatly enhances the rotational diffusion constant of eosin probes attached to the remaining membrane-bound portion.

3.2.3 Physiology of the band 3 protein

It is interesting to consider the role of this protein which is so important to the function of the erythrocyte as to require 1.2×10^6 molecules per cell. Blood not only carries O_2 from lungs to tissues; it also carries CO_2 back from tissues to lungs. The process is believed to occur as follows. During the 0.3–0.7 s that it takes the blood to traverse the tissue capillaries, CO_2 diffuses rapidly into the erythrocyte. It is there converted to H_2CO_3 by cellular carbonic anhydrase. This is deprotonated to HCO_3^-, the H^+ being taken up by buffering groups in the cell (largely on haemoglobin). The rapid exchange of cellular HCO_3^- for plasma Cl^- greatly increases the CO_2-carrying capacity of the blood. This is called the Hamburger shift (Fig. 3.9).

In the lungs the process goes into reverse. The HCO_3^- content of the plasma, comprising 50% of the blood's total carbon dioxide load, can only be expired as gaseous CO_2 after it has entered the red cell (in exchange for Cl^-) and been converted there, first to H_2CO_3 and then to CO_2 and H_2O (by cellular carbonic anhydrase).

Why is there no carbonic anhydrase in the plasma, you may ask. But then you could also ask, why is there no haemoglobin? However such considerations are outside the scope of this book.

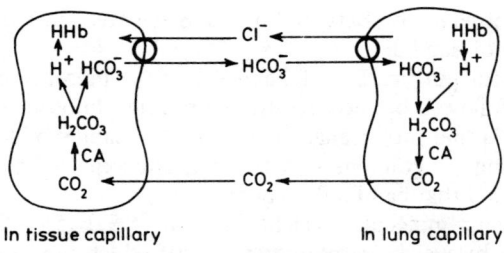

Fig. 3.9 The Hamburger shift. The red blood cell takes up CO_2 in the tissue capillaries and gives it out in the lungs.

3.3 The glucose carrier of the human erythrocyte

There have been a great many papers dealing with the kinetics of glucose transport in the erythrocyte since Widdas postulated such a carrier in 1952, and a similar number of conflicting theories and models. By comparison, the study of the chemistry of the glucose carrier is a recent development, stemming from its successful labelling by Batt, Abbot and Schachter and its functional reconstitution by Kasahara and Hinkle, both in 1976 [35,36]. (For a recent review see [37].)

For a few years glucose transport activity was wrongly attributed to a band 3 component. There is now no doubt that the carrier is a component of the very much fainter band 4.5. The mistake arose in the following way. Cytochalasin B (Fig. 3.10) competitively inhibits glucose transport with a K_I of 1.3×10^{-7} M. It was also found to bind reversibly to band 3. (The small amount bound to band 4.5 was overlooked.) Kasahara and Hinkle, at about the same time, reconstituted functional-carrier into sonicated liposomes using detergent-solubilized proteins including 'band 3 and other minor proteins'. It was further found that proteolytic cleavage of band 3 produced material that ran in SDS electrophoresis as band 4.5, and it was therefore possible to argue that the glucose-inhibitable maleimide labelling

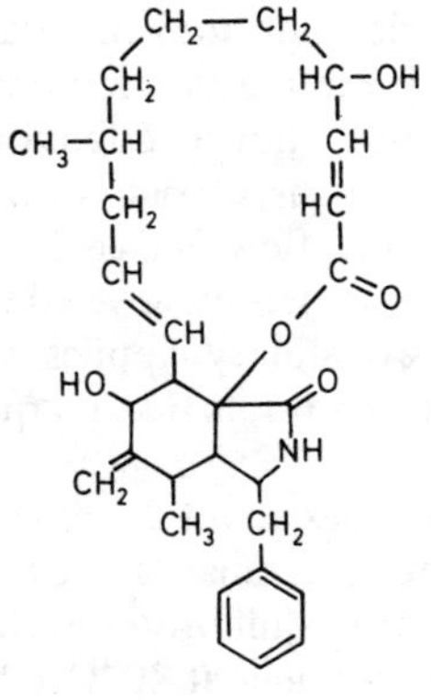

Fig. 3.10 Cytochalasin B.

of band 4.5 observed by Batt, Abbot and Schachter was a result of proteolytic cleavage of band 3.

More careful purification of band 4.5 in the presence of inhibitors of proteolysis, followed by reconstitution, both into liposomes in the manner of Hinkle [38] and into planar bilayers in the manner of Jones [39], confirms that band 4.5 contains a protein with the expected glucose transport specificities, and that band 3 does not.

The apparent molecular weight of band 4.5 is 55 000. The protein contains 15% by weight carbohydrate, all of which can be oxidized in the intact cell by extracellular galactose oxidase. This confirms that the galactosyl residues are exposed only on the outer face of the cell. (Most eucaryotic membrane proteins are glycosylated on their outer face.) After the removal of this carbohydrate the protein runs in sodium dodecylsulphate polyacrylamide gel electrophoresis with an apparent molecular weight of 46 000. There is no evidence that the functional unit is other than the monomer. Nor does cross-linking reveal dimers, though inactivation with high energy electrons indicates a target with a volume equivalent to 200 000 M_r, perhaps a result of temporary and insignificant aggregation.

The asymmetric distribution of carbohydrate more-or-less rules out the tumbling type of 'mobile carrier' (Fig. 2.3), but the glucose carrier, like the anion-exchange carrier, behaves as a gated-pore in which the active-site is exposed alternately to each face of the membrane, and in that sense conforms to the 'mobile carrier' model. This has been demonstrated as follows [37,40]. It has been shown that cytochalasin B is a competitive inhibitor of glucose transport, but that it only binds on the cytoplasmic face of the carrier protein. It has further been shown that glucose analogues with bulky substituents at the 1 position (e.g. n-propyl-β-D-glucoside) also bind only at that face, while analogues with substituents on the other side of the molecule (e.g. 4,6-O-ethylidene-D-glucose) inhibit only from the outside face. The explanation of this may be that the orientation of the sugar on its way through the pore is preserved (Fig. 3.11). From the concentration dependence of the inhibition of cytochalasin B binding by 4,6-O-ethylidene-D-glucose it could be inferred that only one or other of these could bind to the carrier, not both simultaneously.

Not only is the carrier protein asymmetric in the membrane; a striking feature of this carrier is that, though diffusion is passive and no energy input is involved, net flow on the carrier is asymmetric. Glucose can flow out 10 times faster than it can flow in (see Table 2.4). This might seem to contradict the principle of microscopic reversibility, but it must be remembered that microscopic reversibility applies to systems at equilibrium. Under equilibrium conditions isotopic flux is equally fast in both directions (Table 2.4).

The glucose carrier comprises about 2% of the membrane protein of the red blood cell and there are an estimated 1×10^5 molecules per cell. From that figure and the V_{max} for equilibrium exchange (Table 2.4) it can be calculated that the turnover number at 20°C is 1890 s^{-1} (600 s^{-1} at 7°C). The rates of the net, or 'zero-trans', flows indicate that the reorientation

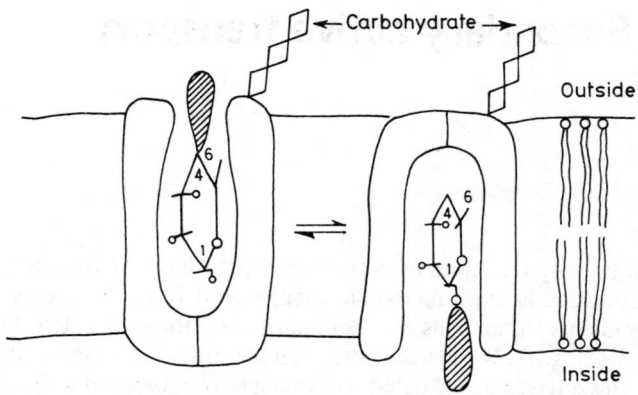

Fig. 3.11 Model of the glucose carrier as a gated-pore with alternating accessibility. Also shown is the asymetric attachment of carbohydrate and how substituents at position 1 on the glucose substrate could prevent access to the binding site from the outside but not from the inside.

of empty carrier from outside to inside occurs at $1027\ s^{-1}$ while that from inside to outside is much slower at $142\ s^{-1}$ ($20°C$) (Fig. 3.12).

As with the anion-exchange carrier it is interesting to enquire why the human erythrocyte requires this very high glucose transport capacity. The answer in this case is less clear. The metabolic needs of the red-blood-cell itself are modest. Glucose can circulate round the body in the plasma without needing access to the internal volume of the red cells, though with an haemotocrit of 0.4, such access would certainly increase the glucose carrying capacity of the blood. In the adult animal this high capacity, asymmetric, glucose carrier system is restricted to the primates; other animals such as the pig and guinea pig, posses it only in the foetus. This might be a clue. We might tentatively suggest that its function is to deliver glucose to the brain. In which case, the very low Sen–Widdas K_m (Table 2.4) is physiologically relevant, for that indicates that, as the plasma glucose level in the cerebral capillaries falls towards 2.4 mM, the rapid net efflux of glucose will begin to occur. At higher extracellular glucose levels, net efflux will be inhibited.

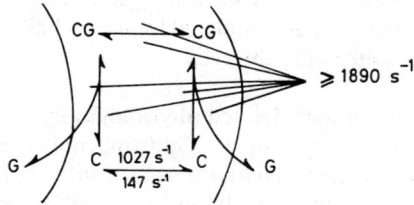

Fig. 3.12 The kinetics of glucose transport in human erythrocytes.

4 Secondary active transport

There are many examples of active transport where the free energy is provided not directly from metabolic changes, but from the energy stored in ionic gradients. In animals Na^+ is in most cases the driving ion. In bacteria and plants it is H^+. While many such coupled transport systems are known, few of these have been isolated and characterized biochemically, and none of them have been fully elucidated. The subject of secondary active transport has been thoroughly reviewed recently [41,42]. I shall confine myself here to a single example; the lactose-proton symport of *Escherichia coli*. This system exemplifies most of the characteristics of symport systems and at the same time is uniquely well characterized at the molecular and genetic levels.

The lactose transport system of *E. coli* has a distinguished history. As part of the lactose operon it contributed to the Nobel prize of Jacob and Monod. As a proton-coupled symport it contributed to the Nobel prize of Peter Mitchell. It was the first carrier protein to be identified radiochemically on an electrophoresis gel (Fox and Kennedy [43]), the first symport shown to be electrogenic and coupled stoichiometrically to proton flux (West [44], West and Mitchell [45,46]), the first to be cloned (Teather et al. [47]).

4.1 Lactose permease defined genetically

The y-gene of the lactose operon, distal to the z-gene (β-galactosidase) and proximal to the a-gene (thiogalactoside transacetylase) was shown to direct the synthesis of a protein responsible for the transport of β-galactoside across the cell membrane [48]. The β-galactosidase enzyme was cryptic ('hidden') in strains with a mutation in the y-gene, and could only act on its extracellular substrate if the cell membrane was physically broken, or lysed with detergent. In metabolizing cells non-hydrolysable thiogalactoside substrates were found to be accumulated by up to 2000-fold, and to remain unmodified in free solution in the cells.

4.2 Lactose-proton symport defined physiologically

The proton-conducting uncoupler dinitrophenol causes not only the inhibition of active uptake but more significantly the release of already accumulated material. Mitchell, the only person at that time (1961) to appreciate the function of dinitrophenol, pointed out that this last observation would be consistent with a model (see Fig. 4.1) in which the proton motive force was responsible for holding in the accumulated galactoside [49]. It was however consistent with other more widely held theories

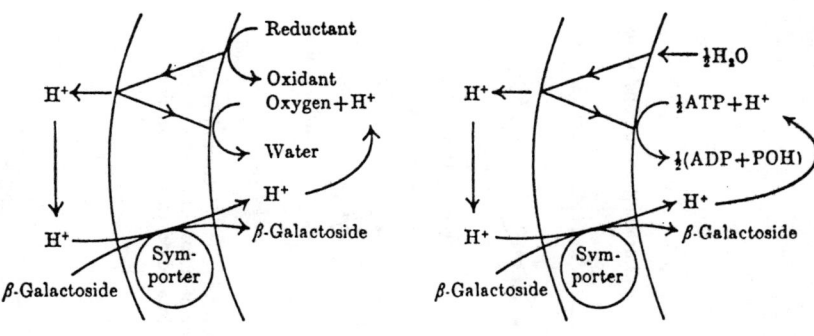

(a) β-Galactoside translocation coupled to respiration

(b) β-Galactoside translocation coupled to ATP hydrolysis

Fig. 4.1 Suggested H^+-symport mechanism of β-galactoside uptake in *E. coli* (after [49]): (a) when the proton current is due to respiration; (b) when the proton current is due to ATP hydrolysis. These diagrams are intended to show the mutual dependence of the flows but not the stoichiometric relationships between them.

about the formation and role of ATP and Mitchell's proposal was ignored for several years.

Pavlasova and Harold [50] confirmed that the effect of uncoupler was not via a lowering of ATP level. This was hardly affected at an uncoupler concentration that abolished accumulation of galactoside. Then West [44] showed that the inflow of lactose into cells could drive the inflow of H^+ ions, confirming that the flux of galactoside and of protons on the carrier are mechanically coupled.

The movement of H^+ ions could be detected as an alkalification of the extracellular medium using a conventional glass electrode. Of course, in the steady state there is no net H^+ inflow. It was important to abolish or minimize the metabolically driven outward pumping of H^+ ions (see Fig. 4.1). That was achieved by doing the experiment under an atmosphere of nitrogen (to stop respiratory H^+ extrusion) and in the presence of iodoacetate (to stop glycolytic production of ATP). The inflow of protons could be shown not only by detecting an alkalification of the external medium but also by detecting an acidification of the cytoplasm (by the consequent accumulation of a radioactive weak acid) [51]. This was important as it confirmed that the proton was indeed translocated into the cells and not just consumed by some metabolic process.

When the cells were made permeable to K^+ ions by adding the K^+-specific ionophone, valinomycin, it was possible, by recording simultaneously with K^+- and H^+-sensitive glass electrodes, to show that when 1 nmole of H^+ entered these cells 1 nmole of K^+ left the cells. That is to say, the inflow of H^+ was electrogenic. The positive potential building up inside the cells under these experimental conditions drove the K^+ out. The inflow of galactoside could be quantitated by using ^{14}C-lactose and was shown to be equal at least at early time points, to the amount of H^+-ion translocated

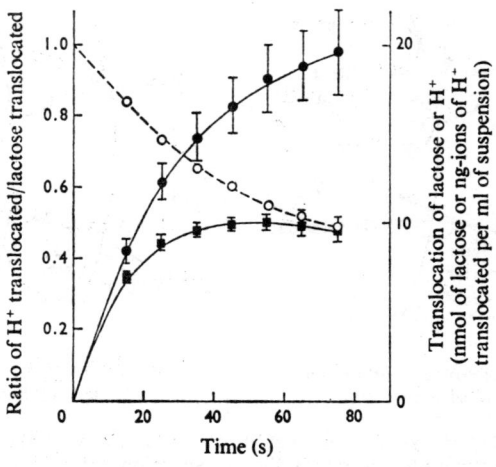

Fig. 4.2 The 1:1 stoichiometry of H^+ and lactose uptake by *E. coli* at pH 7.0.

[46] (Fig. 4.2). This 1:1 stoichiometry has recently been confirmed [52,53] over the pH range 5.8–7.8 and in the presence and absence of a proton motive force (pH gradient or electric field). It is still possible that at high galactoside concentration the carrier might catalyse the inflow of galactoside without H^+ and that has still to be tested.

4.3 The lactose carrier protein

Transport is inhibited by mercurial compounds and other thiol-reagents such as *N*-ethylmaleimide. When the active-site is occupied by certain substrates (thiodigalactoside is particularly effective) reaction with *N*-ethylmaleimide is very considerably slowed ('substrate protection' of the active-site thiol). As the β-galactoside binds tightly and specifically only to the active site of this carrier, all the other thiol groups in membrane proteins are unprotected. Fox and Kennedy [43] cleverly exploited that fact. They reacted the unprotected thiols with ^{12}C-*N*-ethylmaleimide, removed the substrate and reacted the now uniquely reactive –SH group at the active site with ^{14}C-*N*-ethylmaleimide. Membranes so treated were dissolved and electrophoresed in sodium dodecylsulphate, and showed a single labelled band with an apparent molecular weight of 30 000 [54]. I stress the word apparent, for certain types of protein (e.g. glycoproteins and very hydrophobic proteins) are known to run anomalously in that electrophoretic system.

It is estimated that there are about 3×10^3 copies of this carrier protein per cell; too few to visualize with Coomassie stain. For 10 years all attempts to follow up the work of Kennedy were unsuccessful. In a fruitful collaboration between the laboratories of Müller-Hill and Overath (both in Germany) the DNA of the *lac* operon was cloned in a multicopy plasmid. A derivative of that plasmid was selected in which almost all the z-gene and a-gene DNA was deleted. Induction of cells bearing this plasmid produces between three and ten times as much lactose carrier activity and sufficient carrier protein is now present in the membrane (15 times as much as the wild type) to give a detectable spot on polyacrylamide gels [47].

Subjecting the cloned DNA to the rapid sequencing techniques of Maxam and Gilbert and of Sanger has recently provided the DNA sequence of this fragment of the *lac* operon [55]. This DNA was also used to direct *in vitro* protein synthesis. Automated Edman sequencing of this material, and also of the material synthesized *in vivo*, yielded the first few amino acids of the y-gene product and confirmed the reading frame for translation of the DNA sequence [56]. Carboxyl-terminus determination confirmed that the reading frame was still correct through to the end of the molecule. So we now have the complete amino acid sequence of the lactose carrier (Fig. 4.3).

The sequence deduced from the DNA contains 417 amino acids and has a molecular weight of 46 504. Rechecking the mobility in SDS-polyacrylamide gel electrophoresis confirmed an apparent molecular weight of 30 000 in gels with a low percentage of acrylamide (7–10%) but the estimate approached M_r 46 000 in high percentage gels (22%). The protein is unusually hydrophobic (71% of the amino acid residues are hydrophobic) and it seems likely that the anomalous mobility is due to excessive binding of dodecylsulphate [57].

The Chou and Fassman rules that give a reasonably good prediction of the secondary structure of soluble proteins do not work well with membrane proteins. Nevertheless, a scrutiny of the amino acid sequence of lactose permease reveals some interesting features. If a plot (Fig. 4.3a) is made by assigning a hydrophobicity to each amino acid side chain (Table 4.1) through the molecule from the N-terminus it is clear that there are 13 or 14 definite blocks of hydrophobic amino acids interspersed with regions containing more hydrophilic and charged residues (Fig. 4.3b). The first seven, and the last two, of these hydrophobic blocks are particularly clear, each comprising some 20–23 amino acids and therefore long enough to span the phospholipid bilayer as α-helix.

It will be noticed that there are eight cysteine residues, one of which must be the one that reacts with *N*-ethylmaleimide only in the absence of substrate. This 'active-site' thiol has been identified rather neatly by Beyreuther [58]. After specific labelling in the Kennedy manner Beyreuther treated the whole membrane with trypsin and examined the mixture of peptides with an automated Edman sequenator. The emergence of a ^{14}C-labelled cysteinyl derivative after four cycles of N-terminal hydrolysis unambiguously indicates cysteine 148, as you will see by inspecting the sequence (Fig. 4.3b).

It is too early to say anything about whether the permease acts as a dimer, except that dimethylsuberimidate did not result in interchain cross-linking. Nor is it possible to exclude a rotation of the molecule across the plane of the membrane during transport, though such a rotation seems very unlikely.

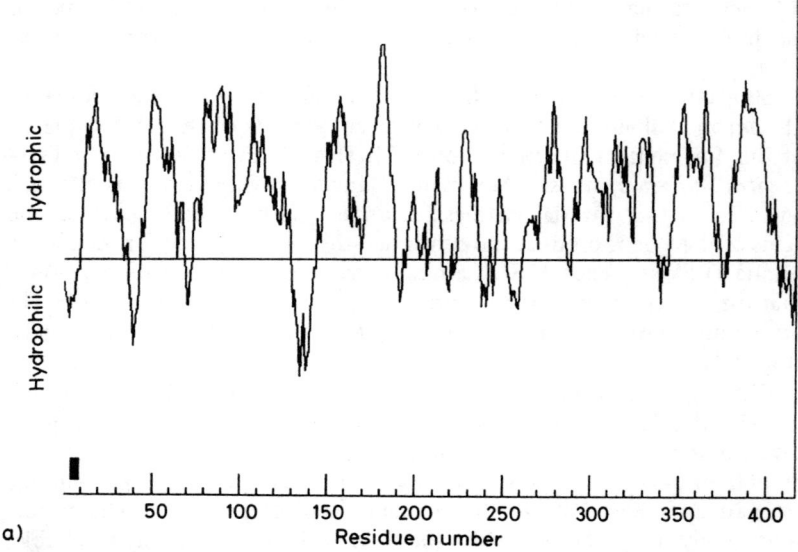

Fig. 4.3 (a) Computed hydrophobicity of the peptide chain of the lactose carrier of *E. coli*.

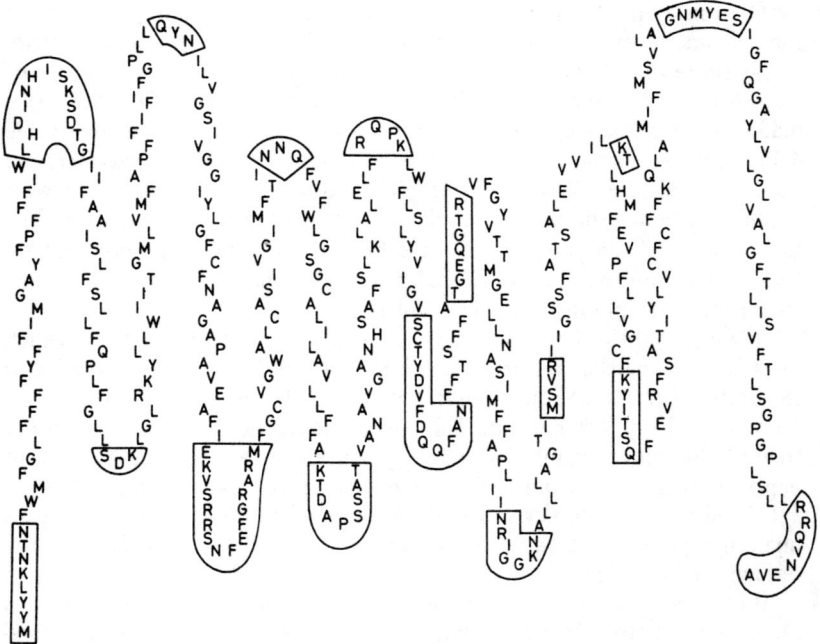

Fig. 4.3 (b) Complete amino acid sequence (single-letter code) showing one possible way in which it might be looped across the membrane. (Hydrophobic segments are boxed.)

42

Table 4.1 Estimated free energy of transfer (ΔG_{tr}) of an amino acid residue from a random coil conformation in water to an α-helical conformation in a lipophilic phase. It is assumed that each H-bond broken requires 10.5 kJ mol^{-1}. It is also assumed that charged sidechains must be neutralized by protonation (Asp, Glu, Cys, Tyr) or deprotonation (Arg, Lys, His).

Residue		Contributions to ΔG_{tr} (kJ mol^{-1})			ΔG_{tr}
3-letter code	1-letter code	hydrophobic	H bond	charge	
Gly	G	7.85			− 7.85
Ala	A	12.04			−12.04
Val	V	16.22			−16.22
Leu	L	17.79			−17.79
Ile	I	−18.32			−18.32
Phe	F	−21.98			−21.98
Tyr	Y	−24.07	+10.50	+12.06	− 1.51
Trp	W	−26.69	+10.50		−16.19
Ser	S	−12.04	+10.50		− 1.54
Thr	T	−14.65	+10.50		− 4.15
Cys	C	−14.13	+10.50	+ 7.58	+ 3.95
Met	M	−19.36	+10.50		− 8.86
Asn	N	−16.75	+21.00		+ 4.25
Gln	Q	−18.84	+21.00		+ 2.16
Pro	P	−15.18	+21.00		+ 5.82
Asp	D	−15.70	+21.00	+17.92	+23.22
Glu	E	−19.89	+21.00	+15.70	+16.81
Lys	K	−20.93	+10.50	+20.14	+ 9.71
Arg	R	−23.55	+31.50	+31.28	+39.23
His	H	−20.41	+21.00	+ 5.69	+ 6.28

4.4 The mechanism of lactose-proton symport

Though there is as yet no final and definite answer to the question how is the proton flux coupled to the lactose flux, it is interesting to see how we can best frame the question in order to examine it. The next few paragraphs are therefore more theoretical or analytical than the foregoing sections.

4.4.1 Thermodynamic approach

Thermodynamic statements, even if they are not obvious, are always clear and certain. Let us first recall Mitchell's equation

$$\Delta p = \Delta \psi - 2.3 \frac{RT}{F} \Delta \text{pH} \qquad (4.1)$$

Where Δp is the proton motive force, $\Delta \psi$ the membrane potential, ΔpH the pH gradient and F is Faraday's constant. (In every case we shall take the quantity in the inner aqueous phase and subtract from it the quantity in the outer reference phase.) This may be written in electrical units as in

Equations 4.1 and 4.2:

$$\frac{\Delta\bar{\mu}_{H^+}}{F} = \Delta\psi - \frac{2.3RT}{F}\Delta pH \qquad (4.2)$$

or the more usual work units (Equation 4.3).

$$\Delta\bar{\mu}_{H^+} = F\Delta\psi - 2.3RT\,\Delta pH \qquad (4.3)$$

The negative sign is merely because pH is defined as $-\log[H^+]$, so Equation 4.1 is often written

$$\Delta p = \Delta\psi + Z\Delta pH \qquad (4.4)$$

where $Z = -2.3RT/F$ and is -58 mV at room temperature. The symbol μ_i signifies the chemical potential of a species i, $\bar{\mu}_i$ the electrochemical potential.

If the flux of one molecule of lactose is mechanically linked to the flux of one proton and if there are no leaks of lactose or protons, then at equilibrium we can write:

$$\Delta\bar{\mu}_{H^+} = -\Delta\bar{\mu}\text{ lactose} \qquad (4.5)$$

This equation is quantitatively obeyed over a wide range of lactose and H^+ concentrations and values for the membrane potential. Apparent deviations at low values of membrane potential could well be due to difficulties in making the measurements. Deviations at high lactose concentration might be expected if at such concentration lactose *could* travel without a proton. This has not yet been conclusively demonstrated.

Note that lactose is uncharged, but that even when there is no pH gradient ($\Delta pH = 0$) lactose would be accumulated by a membrane potential to an equilibrium position such that

$$F\Delta\psi = -\Delta\bar{\mu}\text{ lactose} \qquad (4.6)$$

This has been quantitatively confirmed.

The strength of thermodynamic statements is that they set inviolable limits. For example, thermodynamic arguments can distinguish between lactose travelling with $1H^+$ or $2H^+$, provided that the measurements of $\Delta\psi$, ΔpH and $\Delta\mu_{lac}$ can be made sufficiently accurately, and provided equilibrium is achieved. The weakness is that the thermodynamics can tell us nothing about mechanism; all possible mechanisms will obey the same overall thermodynamic equation.

4.4.2 Kinetic approach

We can perhaps expand a little and show some of the consequences of those thermodynamic arguments. For the carrier to work efficiently and build up a large galactoside concentration gradient there must be very little slip. The carrier must be more-or-less unable to carry galactoside without also carrying H^+. It could conceivably translocate galactoside through half of a transport cycle, but not through a whole cycle. In order not to dissipate energy it seems highly likely that the carrier will likewise be unable (or very nearly so) to carry H^+ without also carrying galactoside. An interesting topic of current research is to determine how closely these two expectations are fulfilled [59,60].

The order of substrate binding is not yet universally agreed; protons first, galactoside first or random order? Let us examine just one of these and choose the one that makes the fewest assumptions yet seems to explain all the data, i.e. the random order [61] (Fig. 4.4).

Suppose that a site on the carrier (C) can bind H^+(H) from the extracellular medium. Suppose that galactoside (S) discriminates between the two forms (C and CH) such that it binds preferentially to CH (except at high galactoside concentration). It is clear that the binding of galactoside changes the carrier in some way, presumably by changing its conformation, because the translocation probabilities of the CH and CHS forms are very different (see above). A 'translocation' step now takes place such that subsequent unbinding to the outside is no longer possible, but unbinding on the inside is possible. (This is the defining feature of 'mobile carrier' models — see Chapter 2.) It is not obvious *a priori* that this 'translocational' conformation change need be either faster or slower than those induced by H^+ binding or by galactoside binding and the temptation to assume that it is the rate-limiting step should be avoided.

The thermodynamic effect of the pH gradient is taken care of by treating H^+ as one of the two substrates. The effect of the electric field, as

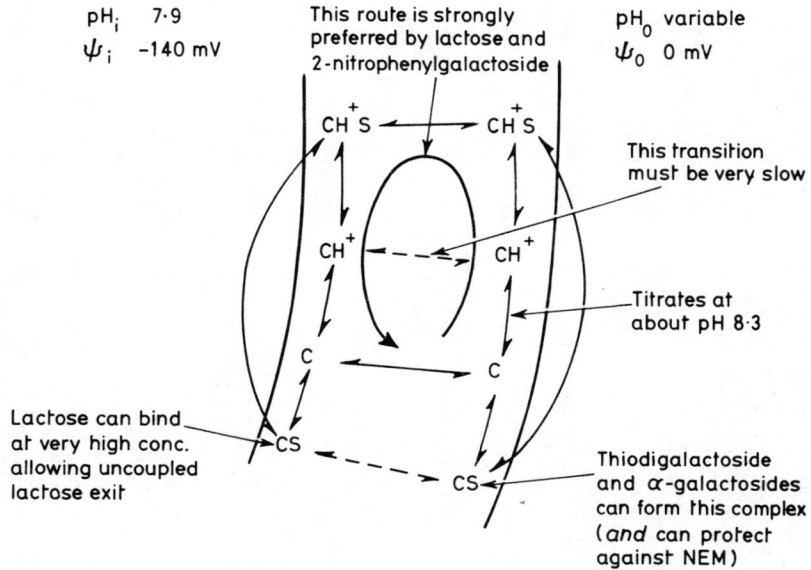

Fig. 4.4 A summary of current information on the kinetics of the β-galactoside carrier of *E. coli*.

explained in Chapter 2, must reside in its effects on H^+ concentrations or on one (or several) of the rate constants in the scheme. At present it looks as though it is the $C_i \rightleftharpoons C_o$ transformation that is affected by $\Delta\psi$; membrane potential (inside negative) speeding up the forward reaction and slowing down the reverse reaction.

It may well prove difficult to demonstrate such a conformational change, though in the case of the $(N^+ + K^+)$-ATPase an analogous conformational difference between two forms of the transport enzyme has been successfully demonstrated (see Chapter 5).

The ultimately satisfying description of coupled transport would be: to show which residues on the carrier bind galactoside and which bind protons; to show the conformational consequences of such binding; and to determine the channel through which the galactoside (and the proton) passes.

5 Primary active transport systems

There are three important and easily distinguishable types of primary active transport systems: (a) ATPases, i.e. membrane-bound enzymes that couple translocation to the hydrolysis of ATP, (b) redox reactions where the effective translocation of protons result from the positioning on either side of a membrane of successive redox reactions, (c) the unique light-driven ion-pumps of certain halophilic bacteria (e.g. *Halobacterium halobium*).

Most ATPases are ion-translocating (H^+, Ca^{2+}, K^+, etc.) though it is possible that current work on bacteria will reveal a number of specific sugar-translocations and maybe also amino-acid-translocating ATPases. In this chapter, of the ATPases, we shall consider only the $(Na^+ + K^+)$-ATPase of animal cells. Of the light driven pumps, we shall discuss the proton-translocating light-driven, bacteriorhodopsin of *H. halobium*. No further reference will be made to the important proton-translocating redox reactions of photosynthesis and oxidative phosphorylation or to the proton-translocating ATPase of bacteria, mitochondria and chloroplasts (see C. W. Jones [102]).

5.1 $(Na^+ + K^+)$-ATPase

This enzyme, though postulated earlier, was first described by Skou [62]. A great deal is now known about certain aspects of its biochemistry, about the chemical changes occurring during the transport cycle and the kinetics of those changes. Relatively little is known about the primary, secondary and tertiary structure of the protein, though rapid progress is currently being made on those questions as well.

The literature on this enzyme is vast and for this three reasons spring to mind: (1) the enzyme plays a fundamental role in the physiology of animal cells; (2) it can be studied in a great variety of different organisms and different tissues; (3) the enzyme is incompletely understood, much of our information is merely suggestive and some of it is even contradictory. The physiology of $(Na^+ + K^+)$-ATPase is largely outside the scope of the present book though a brief discussion of physiology is essential (Section 5.1.8). As for the inconclusive nature of much of the information on the enzyme and the range of different enzymes being studied, these present problems for the student who wishes to grasp the essentials succinctly. In the last year or two a degree of concensus has emerged and a model can now be discussed that satisfactorily incorporates a great deal of the detailed experimental information, though on many points an alternative interpretation is favoured by some workers.

5.1.1 Purification

The $(Na^+ + K^+)$-ATPase is best prepared from a tissue in which it is abundant, i.e. in which ion pumping is an important function. Two such sources are brain 'microsomal' membranes and kidney medulla. Of the many possible tissues and extraction procedures I shall concentrate on the procedure of Jørgensen [63].

Jørgensen prepares the enzyme from the outer renal medulla of either rabbit or pig (the major portions of the ascending and descending tubules of the loop of Henle). The preparation depends on careful addition of the right amount of the powerful but denaturing anionic detergent, sodium dodecyl sulphate (SDS) in the presence of 3.3 mM ATP. It is possible that the ATP protects the enzyme under these conditions. At any rate essentially all *other* proteins can be dissolved. A single isopycnic-zonal centrifugation in a sucrose gradient suffices to separate the $(Na^+ + K^+)$-ATPase still bound to membrane (density 1.117 g ml^{-1}) from the dissolved proteins. This preparation yields an almost pure protein plus an equal weight of phospholipid with an SDS content of less than 0.5%.

Another interesting purification [64] exploits the fact that both subunits of the $(Na^+ + K^+)$-ATPase are glycosylated and uses an affinity column bearing immobilized concanavalin A, the plant lectin with a very high affinity for mannosyl residues. The purest preparations have a specific activity of $20-40$ μmol min^{-1} (mg protein)$^{-1}$.

Electrophoresis under denaturing conditions in SDS shows two bands of protein stainable with Coomassie Blue — a dense one at around $100\,000$ M_r and a fainter one at around $50\,000$ M_r. When allowance is made for the smaller size of the latter and for the fact that it is a sialoglycoprotein and thus binds unduely little SDS and Coomassie Blue it is estimated that there are two types of subunit (α and β) in equal molar ratio, with relative molecular weights of $100\,000$ and $40\,000$ respectively considering only the protein component. (The α-subunit $90\,000-106\,000$ and the β-subunit $32\,000-65\,000$ depending on source [66].)

The smaller (β) subunit is probably not a contaminant. Antibodies prepared to purified β-subunit inhibit the enzyme when applied extracellularly. This subunit is quite heavily glycosylated (about 28% w/w), so it is thought that at least part must be exposed on the outer face, corroborating the antibody result. The larger (α) subunit is often referred to as the catalytic subunit. On incubation of native enzyme with [^{32}P]-ATP under the right conditions (e.g. 100 mM Na^+, 3 mM Mg^{2+}, but no K^+) the active site of the enzyme can be phosphorylated. Subsequent analysis by SDS polyacrylamide gel electrophoresis shows that it is the larger subunit that carries the ^{32}P label. The α-subunit quite clearly spans the membrane from the following considerations. Red blood cell membranes can be prepared outside-out, inside-out and with both sides exposed to the medium. ATP clearly reacts on the cytoplasmic side and the non-penetrating inhibitor ouabain binds tightly to the α-subunit from the outside. If the purified enzyme is reconstituted in vesicles of purified phosphatidyl choline it adopts both inside-out and outside-out orientation with about equal

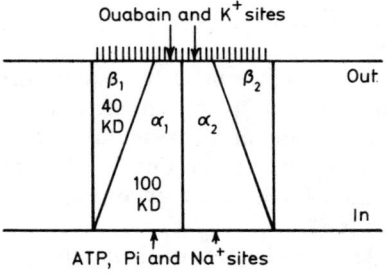

Fig. 5.1 $(Na^+ + K^+)$-ATPase.

frequency. However, external ATP (non-penetrating) only reaches the active site of those pumps with inside-out orientation. The resulting activity is not inhibited by ouabain, unless the ouabain is incorporated inside the vesicle (Fig. 5.1).

5.1.2 Quaternary structure

The quaternary structure of the active enzyme is still uncertain and a subject of continued debate, but the bulk of the evidence favours an $\alpha_2 \beta_2$ tetramer.

One of the earliest indications for such a structure came from radiation inactivation studies which indicate a target size that would correspond to a relative molecular weight of 200 000–500 000 if the protein were a sphere. Jørgensen and colleagues [65] have applied the techniques of freeze-fracture to membrane preparations of the enzyme and on high resolution electron micrographs they have visualized particles with a diameter of 9 nm. That also suggests a molecular weight of around 200 000.

Gel filtration of enzyme dissolved in a non-ionic, non-dissociating detergent (Lubrol) indicated a molecular weight of 500 000, but it was difficult to quantify the contribution of Lubrol to the volume of the micelles.

The subunit molecular weights of around 100 000 and 40 000 obtained by SDS-electrophoresis have been confirmed by sedimentation equilibrium centrifugation [66]. From such subunits it would be possible to form molecules, falling within the very broad weight range indicated, with the composition $\alpha_1 \beta_1$, $\alpha_1 \beta_2$, $\alpha_2 \beta_2$, $\alpha_2 \beta_4$, $\alpha_3 \beta_3$, $\alpha_4 \beta_2$, etc. However, the measured weight ratio of the subunits (α/β) from gels and from amino acid composition is around 2.3 so only $\alpha_1 \beta_1$, $\alpha_2 \beta_2$ and $\alpha_3 \beta_3$ are candidates.

At 0°C and in the absence of divalent cation the enzyme will bind ATP with high affinity, but without hydrolysing it. It is therefore possible to determine the weight of enzyme protein that binds one mole of the inhibitor, ouabain, or one mole of the substrate ATP. In both cases a single mole of binding site was present per 278 000 g protein. That confirms very closely the proposed $\alpha_2 \beta_2$ structure and the molecular weight of the functional molecule, but it raises another problem: if there are two α-subunits why did not two ATP or two ouabain molecules bind? We shall return to that question later.

These measurements therefore indicate an $\alpha_2\beta_2$ structure. Direct conformation has been obtained from cross-linking studies. Dimethylsuberimidate readily cross-links α-subunit to β-subunit as a dimer; not however β-subunit to itself. On the other hand, Cu^{2+} in the presence of o-phenanthroline can cause cross-linking between α-subunits by oxidizing –SH groups to S–S bridges [63].

The proposed quaternary structure of these subunits and their insertion in the membrane is shown in Fig. 5.1.

5.1.3 Phosphorylated enzyme intermediates

This enzyme is called the $(Na^+ + K^+)$-ATPase because ATP hydrolysis is greatly stimulated by the simultaneous presence of Na^+ and K^+; in addition, and in common with most other reactions in which ATP participates, Mg^{2+} is also required (Ca^{2+} is much less effective.) Experiments with intact erythrocytes and resealed ghosts confirmed that K^+ stimulates from the outside (in the concentration range 0.1–1 mM) while Na^+ stimulates from the inside (in the concentration range 30–100 mM).

In the absence of Mg^{2+}, ATP is still bound to the enzyme ($K_D \simeq 0.2\ \mu M$) but it is not hydrolysed, as mentioned above. In the absence of Na^+ (K^+ and Mg^{2+} present) ATP no longer binds to the enzyme with high affinity (K_D rises 1000-fold to 0.5 mM) and ATP hydrolysis is extremely slow.

In the absence of K^+ (Na^+ and Mg^{2+} still present) ATP is still hydrolysed, but again at a very much slower rate. Moreover, if the reaction is quenched with $HClO_4$ under these conditions, the α-subunit of the precipitated enzyme is found to be phosphorylated. When $\gamma[^{32}P]$-ATP was used, and the protein digested with the proteolytic enzyme pronase, a phosphorylated tripeptide was isolated with the sequence:

$$- \text{Ser (or Thr)} - \underset{\underset{^{32}P}{|}}{\text{Asp}} - \text{Lys} -$$

The acid anhydride bond between an aspartyl carboxyl and a phosphate would normally have a high free-energy of hydrolysis. This sequence is particularly interesting because an identical sequence has been found round the phosphorylated aspartyl in the closely related Ca_2^{2+}-ATPase of the sarcoplasmic reticulum.

Quantitative measurements of the amount of phosphorylated enzyme (E–P) during steady-state turnover in the presence of only Na^+ and Mg^{2+}, found one P atom per molecule of enzyme; i.e. one P atom per two α-subunits, if that previous conclusion was correct. On adding K^+ to the incubation, the steady-state level of E–P rapidly falls to near zero.

Two important states of the enzyme can therefore be distinguished: unphosphorylated and phosphorylated (E and E–P). So also can two partial reactions: the binding of ATP with transfer of its terminal phosphate to the enzyme, which requires Na^+ and Mg^{2+}, and the hydrolysis of E–P to E which requires K^+.

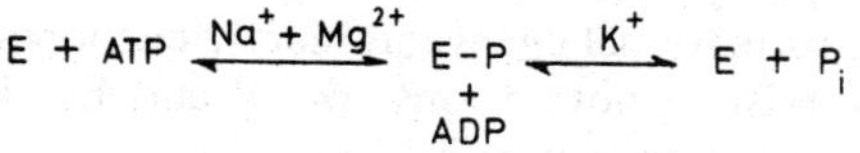

E + ATP
Na⁺+ Mg²⁺
E–P
+
ADP
K⁺
E + Pᵢ

Scheme 5.1

(It has been argued by Skou [67] and others that E–P may not be a true intermediate on the normal catalytic pathway, for it is prominent only in the absence of K^+, when very little (< 4%) ATPase activity is observed. However, this phosphorylated intermediate remains a key element in most current speculation on the action of the enzyme.)

E–P can be dephosphorylated to E either by K^+ as described, or by adding ADP (see Scheme 5.1). However, adding ADP does not dephosphorylate all of E–P. The proportion that is 'sensitive' to ADP increases with increasing Na^+ in the region 10–100 mM, or on adding *N*-ethylmaleimide (NEM). As the fraction of E–P that is 'ADP–sensitive' increases, the fraction that is 'K^+-sensitive' decreases. Post *et al.* [68] suggested the presence of two forms of the E–P intermediate, E_1–P (ADP-sensitive) and E_2–P (K^+-sensitive).

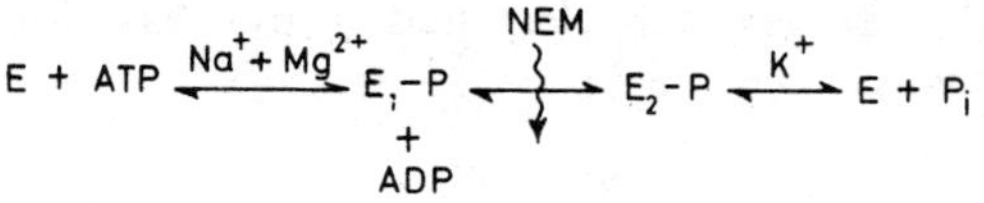

E + ATP
Na⁺+ Mg²⁺
E₁–P
+
ADP
NEM
E₂–P
K⁺
E + Pᵢ

Scheme 5.2

5.1.4 Two conformations of the enzyme: E_1 and E_2

The conclusions that there are two forms of the enzyme, E_1–P and E_2–P, independent of the presence or absence of the phosphate group, that the transition between them can be blocked by *N*-ethylmaleimide, and that the balance between them is determined by Na^+ concentration, suggested the following scheme for the cyclic operation of the pump. It is known as the Post–Albers scheme after its two major originators [68,69].

K⁺
Na⁺
out
E₂– P.K
E₂– P. Na
NEM
Pᵢ
E₂(K)
E₁– P(Na) + ADP
in
E₁ ATP
Na⁺+ Mg²⁺

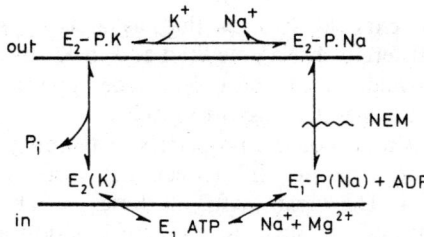

Scheme 5.3

The phosphorylation of the enzyme in the presence of Na^+ and Mg^{2+} (bottom right in Scheme 5.3) is supposed to encourage the conformational

change between E_1 and E_2, between inward facing, Na^+-preferring, carrier and an outward facing, K^+-preferring, conformation. It is worth emphasizing that the phosphorylation of the enzyme does not, in this scheme, constitute the conformational change; it merely encourages it. Phosphorylated enzyme can exist in both forms, E_1–P and E_2–P, as can unphosphorylated enzyme (bottom left in Scheme 5.3).

There is considerable evidence for such a scheme. It is interesting that antibodies to the α-subunit, which only inhibit from the inside surface incidentally, inhibit ATPase activity but do not inhibit the $Na^+ + Mg^{2+}$ stimulated phosphorylation of the enzyme or the $[^{14}C]$-ATP $\rightleftharpoons$ ADP isotope exchange. It is clear that the antibodies are not preventing access of ATP to its binding site; but they could be preventing the $E_1 \rightarrow E_2$ conformational change, by trapping the enzyme in the E_1 conformation.

The most conclusive evidence for the two conformations has come from Jørgensen's laboratory [63]. If isolated enzyme is suspended in 150 mM KCl and treated with trypsin, there is a monophasic destruction of all enzymic activity with a $t_{1/2}$ of 20–30 min. If the enzyme is instead suspended in 150 mM NaCl, there is an initial very rapid ($t_{1/2}$ = 5 min.) loss of part of the ATPase activity, followed by a 20 times slower destruction of the remainder.

If the cleavage products obtained in KCl and NaCl are compared in SDS-polyacrylamide gels it is clear that trypsin has cleaved at different places in the two salts (Fig. 5.2).

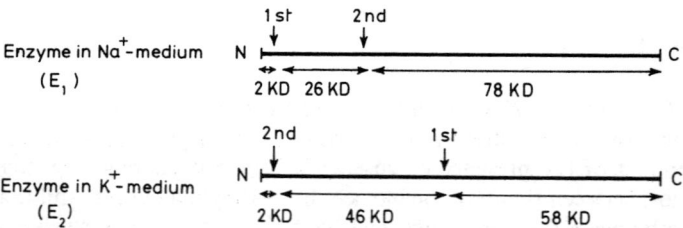

Fig. 5.2 Trypsin cleavage sites on α-subunit. (The rapid first cleavage in NaCl does not detectably change electrophoretic mobility but changes the N-terminus.)

These experiments clearly show that, as far as trypsin is concerned, the enzyme looks different in Na^+-medium and in K^+-medium. Furthermore, this approach would seem to have very wide application for investigating conformational changes in transport proteins.

Further investigation of the properties of the enzyme in NaCl and KCl revealed a number of other differences that correlated with the trypsin-cleavage patterns. The most useful of these are tabulated in Table 5.1. (Remember that Na^+ favours the E_1 conformation and that K^+ favours the E_2 conformation.)

5.1.5 Free-energy changes

The most efficient way to do work is to do it reversibly, that is to say

Table 5.1 Properties of E_1 and E_2 conformations of $(Na^+ + K^+)$-ATPase

E_1	E_2
Prevails in Na^+-medium	Prevails in K^+-medium
Na^+ binds more strongly than K^+	K^+ binds more strongly than Na^+
E_1–P sensitive to ADP	E_2–P sensitive to K^+
High affinity for nucleotides	Low affinity for nucleotides
Low affinity for vanadate	High affinity for vanadate
Low intrinsic (Trp) fluorescence	High intrinsic fluoresence (103%)
High extrinsic fluorescence from eosin-maleimide- or fluorescein-isothiocyanate-labelled enzyme	Low fluoresence of labelled enzyme (70%)

against a counter force that nearly equals the driving force. In that way a minimum of the available free-energy will be dissipated as heat. It is interesting that oxidative phosphorylation and the $(Na^+ + K^+)$-ATPase both operate in this manner. You will notice that all the arrows in Scheme 5.3 are drawn with a barb at each end; the pump can be reversed, and can make ATP if given the right ion gradients.

Considering the first two lines of Table 5.1, it is clearly no accident that the E_1 form prevails in Na^+-medium while it is also the form with the higher Na^+ affinity. The free-energy trough containing E_1. Na^+ draws both cation and enzyme into this conformation. This introduces the idea of 'reciprocal relations', elaborated by Wyman [70]. If the binding of X to a protein affects the affinity for Y then the binding of Y will affect the affinity for X. Further examples follow.

Considering the 4th line in Table 5.1, one would expect ATP to favour the conversion from $E_2 \to E_1$, and that is indeed found to be the case. In the presence of 100 mM K^+ and no Na^+, high concentrations of ATP (5 mM) give both the fluorescence signals and the trypsin-cleavage patterns expected of the Na^+-form of the enzyme (E_1).

Look now at the fifth line in Table 5.1. The inorganic vanadate ion resembles the pentacovalent intermediate formed during phosphoryl transfer:

$$ADP - O - \overset{\overset{\displaystyle O}{\|}}{\underset{\underset{\displaystyle O\ \ O}{\diagup\!\diagdown}}{P}} - O - Enz$$

It is thought that this is the reason vanadate binds to many phosphoryl transferring enzymes with very high affinity. Well-made enzymes should bind the intermediates of the reactions they catalyse more tightly than substrate or product [71]. Vanadate is reported to bind to two sites per ATPase molecule. At one site it binds tightly (K_D = 4 nM), and competitively with ATP (K_D = 0.5 mM), at the other less tightly (K_D = 500 nM).

At this second site ATP competes much more strongly, suggesting that this is the tight-binding site on the E_1 conformation. Intermediate concentrations of vanadate should therefore bind to E_2 and favour the E_2 form. That in turn should favour K^+ binding over Na^+, which it does, to the extent of a 5-fold increase in K^+-affinity [72].

Ouabain (also called strophantin G), is the most important inhibitor of this enzyme. It is one of a class of compounds called cardiac glycosides because of their effect on the heart. It binds tightly to the enzyme from the extracellular face of the membrane. Binding is enhanced by Na^+, weakened by K^+. Further, ouabain binding in the presence of Mg^{2+} and inorganic phosphate causes the formation of phosphorylated enzyme. It is concluded therefore that ouabain, by binding most strongly to E_1-P, favours the right hand side in the following two equilibria:

$$E_2 \rightleftharpoons E_1$$
$$E + P_i \rightleftharpoons E-P + H_2O$$

5.1.6 Half-of-the sites reactivity

It is not infrequently found with oligomeric enzymes that though two identical subunits are present, only one can be detected by active-site titration. Intersubunit co-operativity ensures that while one subunit is in configuration a, the other is in configuration b, and during the enzymic cycle they alternate, 180° out of phase. This phenomenon is known to enzymologists as 'half-of-the-sites reactivity'.

There are several indications that $(Na^+ + K^+)$-ATPase displays half-of-the-sites reactivity. They are summarized below (See [73] for further information).

(1) There appear to be two α-subunits per molecule yet only one ouabain is bound per molecule, and only one ATP with high affinity and one vanadate with high affinity are bound per molecule.
(2) There is evidence that K^+ and Na^+ sites exist at the same time, not alternating as in Scheme 5.3. Thus α^1 could be in the Na^+-form while α^2 is in the K^+-form.
(3) It has been pointed out [73] that the free-energy profile could be evened out still further if the passage of one subunit through a difficult transition were coupled to that of the other subunit through an excessively favourable step. There is some (rather complex) evidence that this is so.
(4) There is some evidence that digitonin dissociates the two subunits of the enzyme. It also blocks ATPase but leaves some other partial reactions of the pump unaffected.

The above arguments would be greatly weakened if it turned out that the smallest active unit was the $\alpha\beta$ dimer, instead of the $\alpha_2\beta_2$ tetramer as is normally believed.

5.1.7 Single-turnover kinetics

The $(Na^+ + K^+)$-ATPase is unique amongst transport catalysts in the extent of our knowledge of its single-turnover kinetics, largely due to a series of papers by Karlish, Yates and Glynn [74,75]. By exploiting the fluorescence signals mentioned in Table 5.1, and a fluorescent analogue of ATP called formycintriphosphate, it was possible to determine photometrically the rates of many of the transitions in Scheme 5.3. The results will be summarized in terms of the slightly expanded scheme from Karlish et al., [74] (Scheme 5.4).

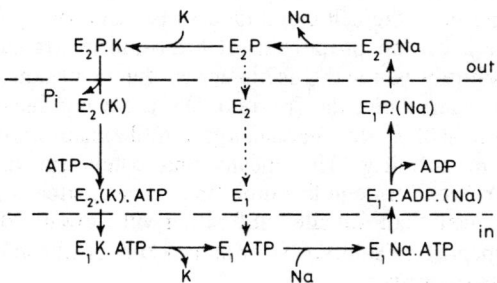

Scheme 5.4

The overall turnover number of the entire enzyme cycle is 205 s^{-1} at 37°C, but around 60 s^{-1} at 20°C. The following data all relate to 20°C. Starting at the bottom right hand corner of Scheme 5.4, the phosphorylation of the enzyme is rapid (183 s^{-1}); the conformational change E_1–P → E_2–P is slower (76 s^{-1}); the exchange of cations is assumed to be fast; the dephosphorylation E_2–P → E_2(K) is very fast (233 s^{-1}). The K$^+$ in this form of the enzyme is shown in brackets to indicate that it is occluded, i.e. it is not exchangeable with isotopes or analogues (Rb$^+$, etc.) from either aqueous phase. The step E_2(K) → E_1 in the absence of ATP is phenomenally slow (0.23 s^{-1}). High ATP concentrations are required to accelerate this step and with saturating ATP (10 mM) and Na$^+$ the rate increases to 54 s^{-1}, but this remains the rate-limiting step of the whole cycle. As remarked above (Section 5.1.5) the free-energy of binding of ATP may be looked on as pulling the enzyme over into the tight-binding E_1 form.

The above measurements relate to the chemical events, not to transport. It is assumed [63,76], but not yet proven, that it is those two slow transitions between the two conformations of the enzyme that constitute the transport step in the context of the gated-pore model.

5.1.8 Physiological role of the (Na$^+$ + K$^+$)-ATPase

Essentially all living cells have a raised K_i^+/Na_i^+ ratio compared with the ratio K_o^+/Na_o^+ (i and o subscripts denote intracellular and extracellular respectively.). Animal cells achieve this disequilibration by the simultaneous pumping out of Na$^+$ and pumping in of K$^+$. In most tissues 3 Na$^+$ ions are pumped out for every 2 K$^+$ ions pumped in and for every ATP hydrolysed, and the pump thus generates an electric field across the cell-membrane; it is said to be *electrogenic*. In some tissues however, e.g. in frog skin epithelium, the pump appears to be *electroneutral*, though this may be due to leaks rather than to the pumping of 2 Na$^+$ or 3 K$^+$.

Besides the maintenance of an elevated cell potassium concentration the pump serves three other important functions.

(1) The pump maintains osmotic equilibrium and prevents the steady inflow of water which, in these fragile cells that are devoid of rigid walls (c.f. bacteria, fungi, and green plants), would cause swelling and eventually bursting. Many animal cells, when deprived of oxygen and therefore of ATP soon develop 'blebs' or 'blisters' where the increased cell water forces the cell-membrane away from the cytoskeleton. This is because a certain proportion of the osmotic strength of the intracellular solution is not diffusible across the membrane, either because it is itself macromolecular (protein, DNA, etc.) or because it is held in the electric field of the fixed charges on the macromolecules to maintain electroneutrality. This means that when the diffusible solutes have distributed to equilibrium there is a greater osmotic strength inside the cell than outside, and water will flow in. Only by the net active transport of solute (Na^+ + anion) out of the cell can continued swelling be prevented.

(2) The pump maintains electrochemical potential gradients, of both K^+ and Na^+ across the cell membrane. These gradients store free-energy and are used in two important ways. (i) Rapid transient opening of passive channels selective for either Na^+ or K^+ can cause rapid depolarization (loss of the trans-membrane electric potential) and repolarization respectively. In some specialized cells, such as those of nerve and muscle, the opening of the channels is under the control of the electric potential, so this wave of depolarization is self-propagating. This constitutes the nerve impulse. (ii) The inwardly directed Na^+ electrochemical potential gradient is used to drive the symport of many small solutes in animal cells and possibly also the antiport of Ca^{2+} from some tissues (though a Ca^{2+}-ATPase is also found).

(3) The pump establishes and maintains an electric potential across the membrane. In reality this is only a trivial corollary of function (2) above, but it is worth treating separately as it is so easily misunderstood. If the pump is electrogenic, pumping 3 Na^+ out for every 2 K^+ pumped in, it is easy to see that the cell interior would go negative (relative to an extracellular reference electrode). If the pump itself were electrically neutral (2 Na^+, 2 K^+ or 3 Na^+, 3 K^+) but if the membrane were differentially permeable by the two ions (as is in fact the case), such that more K^+ *tended* to flow out than Na^+ in, then *net* pumping would once again show an imbalance. The overall effect would be electrogenic even though the pump itself could be electroneutral.

The electric potential that is found across essentially all living membranes will affect the passive concentration distribution across the membrane of all other ionic species. But it is hard to think of cases where the electric potential, as distinct from an ionic electrochemical potential gradient, is itself used for transport, communication, or the storage of free-energy.

The striking exception is the electric organ of the electric eel, torpedo rays, and certain other fish, where the electric potential itself is used for defensive, aggressive or navigational purposes [77].

It would be remarkable if such a crucial enzyme, that regulates cell volume and provides the motive power for solute uptake, solute excretion, volume secretion, electrical signalling, etc., were not regulated in some way by the needs of the cell. Little is known of this aspect of $(Na^+ + K^+)$-ATPase. However, Racker has recently suggested that the enzyme can be phosphorylated by protein kinase [78].

5.2 Bacteriorhodopsin
This protein is well worth looking at in some detail, not because it represents a common type of membrane transport carrier, but because it is unique; it is the only membrane protein for which the secondary and tertiary structure is known. The amino acid sequence is also known, so a very detailed model of this interesting protein can be built, and its manner of functioning guessed at.

5.2.1 Structure
The protein is a single polypeptide chain with a molecular weight of 26 000. The chromophore is the same as that of the visual pigment rhodopsin, namely a retinal group attached by a Schiff's base linkage via its —CHO group to a lysine —NH_2 group of the protein (Fig. 5.3).

Henderson and colleagues [79] developed a novel technique to obtain crystallographic information about this protein in spite of the fact that it cannot be crystallized in the normal 3-dimensional way. It is packed so tightly into the purple patches in the cell membrane that it presents a 2-dimensional regularity in the plane of the membrane. Henderson's group first photographed the electron diffractions obtained from two planes at right angles. To solve the 'phase' problem they took an EM transmission photograph but, as normal EM exposures would completely destroy the sample they used a very low exposure and searched the low-contrast photograph for regularities using a computer. The resulting electron density map that they built up clearly showed seven α-helical rods of protein

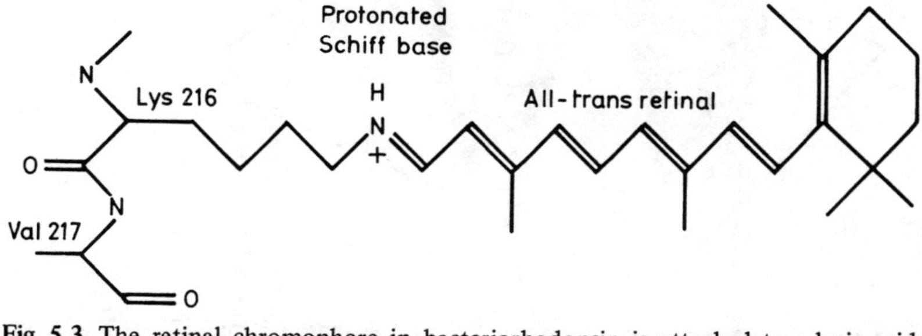

Fig. 5.3 The retinal chromophore in bacteriorhodopsin is attached to a lysine side chain by a protonated Schiff's base.

passing up and down through the membrane, forming a somewhat flattened cylinder with its axis at right angles to the plane of the membrane (Fig. 5.4).

The primary structure has recently been obtained by two groups; that of Khorana in America and of Ovchinnikov in Russia [80,81]. The Russians have also determined the points in the primary structure where proteolytic enzymes can cleave the native protein embedded in the membrane. These are listed in Table 5.2, and unambiguously indicate certain of the points where the peptide chain is exposed to the aqueous media Fig. 5.5.

The amino acid side chain to which the retinal is attached was determined by first reducing the Schiff's base to a stable linkage with borohydride. Subsequent sequencing revealed a modified Lys216. (Earlier suggestions that Lys41 was involved were a result of incomplete knowledge of the sequence.)

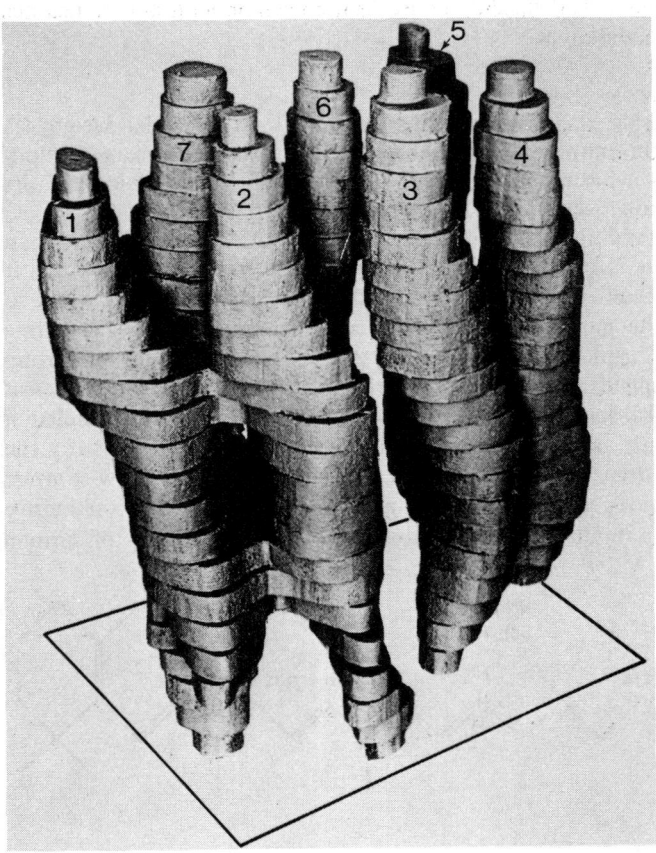

Fig. 5.4 Model of bacteriorhodopsin at 0.7 nm resolution (from [79]). The seven α-helical rods are numbered arbitrarily. Their probable sequence along the peptide chain is 1,76,5,4,3,2 (see [103]).

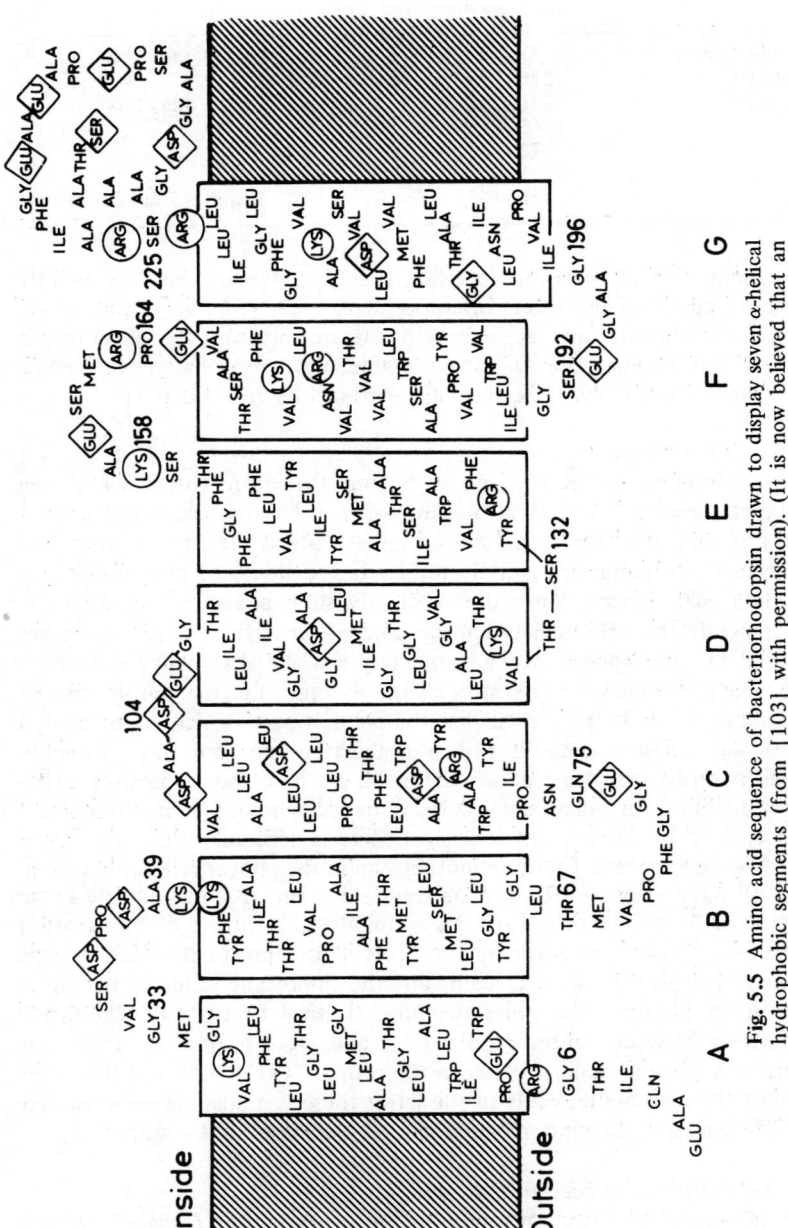

Fig. 5.5 Amino acid sequence of bacteriorhodopsin drawn to display seven α-helical hydrophobic segments (from [103] with permission). (It is now believed that an extra Trp should be inserted at residue 138.)

Table 5.2

Enzyme	Cleavage between residues numbered	Comment
Chymotrypsin	71–72	On the outer face
Trypsin	238–239	
Papain	231–232	On the inner face
	65–66	
	72–73	
	3–4	
	162–163	Requires 7 days digestion

Putting this information together, it is possible to construct a fairly accurate model of the bacteriorhodopsin molecule (Fig. 5.5). It is not yet possible to indicate the precise route of the protons through this molecule, though it is assumed that the protonatable and deprotonatable side chain residues (Lys, Arg, Tyr, Cys, His, Glu, Asp) will be involved [82].

5.2.2 The photocycle

The absorbance spectrum of native bacteriorhodopsin (bR_{570}) has been examined both before a flash of light, when it has an absorbance peak at 570 nm, and at different intervals after a flash. A sequence of modified molecular configurations is indicated by the sequence of new absorbance maxima that appear. These are succinctly summarized in Fig. 5.6. The most significant intermediate appears some 40 μs after the flash and takes several ms to disappear. It has completely lost the absorbance at 570 nm and absorbs somewhat less strongly at 412 nm, i.e. the purple protein becomes yellow. It is believed that the Schiff's base, which is protonated in bR_{570}, is deprotonated in this intermediate designated M_{412}. Roughly synchronously with the appearance of M_{412} there appear protons in the outer medium. An embarrassing detail is that current measurements indicate two ejected H^+ ions at this stage, yet there is definitely only one retinal Schiff's base proton. The rate limiting step in the photocycle is the relaxation of M_{412} back to bR_{570}. (The transient form N_{520} may not be a true intermediate on the path from M_{412} to bR_{570}.) During that relaxation protons are taken up again by the protein to reprotonate those groups from which protons were ejected, but the important point is that these protons come from the inner medium. It must be that the significant difference between conformations L_{550} and O_{640} is that in the former there is a proton channel between the retinal-Lys216 area and the outer face of the membrane, while in the latter there is a channel between that Schiff's base and the inner membrane face. Details are not yet known.

5.2.3 Physiology of bacteriorhodopsin

The net result of continuous illumination of the purple *H. halobium* is to cause a continuous current of H^+ ions to be driven out of the cell. In the steady-state these protons have to return to the cytoplasm, but as the proton-conductance of the membrane is low (or the resistance to proton

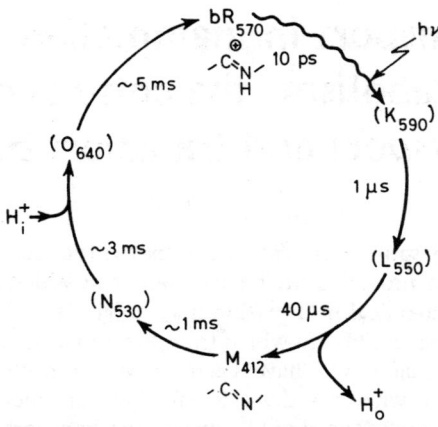

Fig. 5.6 The photoreaction cycle of bacteriorhodopsin (after [105]).

flux is high) there develops a considerable potential drop across the membrane. The passage of H^+ ions down a gradient of $\tilde{\mu}_{H^+}$ (the electrochemical potential of protons) can yield useful work as can the passage of electric current down a voltage gradient. It is, of course, well known now that in bacteria, as in mitochondria and chloroplasts, there is a reversible H^+-translocating ATPase capable of synthesizing ATP from ADP and P_i as protons pass through it into the cytoplasm. The free-energy per mol at which the ATP is made appears to be about twice the free energy yielded by the translocation of H^+ inwards down its electrochemical potential gradient. So it seems that $2H^+$ are required per ATP synthesized.

The purple pigment in the membrane of *H. halobium* develops as the salt pans, in which this organism thrives, dry out and the O_2 content of the saturated salt solution sinks. It has been shown that the light-driven proton pump can supplement the respiratory proton pumps to maintain both the proton motive force necessary for ATP synthesis and the alkalinity of the cytoplasm.

The other striking feature of this organism is the large difference of the electrochemical potential of Na^+ ($\Delta\tilde{\mu}_{Na^+}$; i.e. $\tilde{\mu}$ inside $-$ $\tilde{\mu}$ outside) that it maintains across the plasmalemma.

$$\Delta\tilde{\mu}_{Na^+} = \Delta\Psi + 2.3\,RT \log(Na_i^+/Na_o^+)$$

$$= -180 + 60 \times (-1) \quad mV$$

$$= -240 \text{ mV}$$

K^+, though accumulated 1000-fold, is close to equilibrium in illuminated cells because the membrane potential ($\Delta\Psi$) is about -180 mV (inside relative to outside). Thus, $\Delta\tilde{\mu}_{Na^+}$ is approximately the same as the $\Delta\tilde{\mu}_{H^+}$, and it is possible that it could be maintained by a simple electroneutral (1:1) Na^+/H^+ antiport. However, there appears to be a separate, light driven, pump called halorhodopsin that pumps Na^+, much as bacteriorhodopsin pumps H^+.

There is still much to learn about the physiology of these bacteria.

6 Transport in mammalian metabolism: the control of transport and transport diseases

In this chapter some examples of transport processes in the mammal will be discussed to provide a biological context in which the preceding structural and biochemical information may be viewed and its relevance seen. This is only one context in which transport processes are relevant; illustrations could equally well have been drawn from microbiology or plant physiology, for wherever there is life there are membranes, and where there are membranes there will be membrane transport processes.

One group of topics demonstrates the importance of the control of transport and illustrates the sorts of mechanisms that are involved. Another group of topics illustrates how certain diseases are a result of transport failures, or failures of the control mechanisms. These groups overlap.

6.1 Gastric acid secretion

One of the most striking concentration gradients established in the mammal is the pH gradient across the membrane of the parietal cells (also called oxyntic cells) of the stomach. H^+-ions can be pumped from the cytoplasm (pH 7.3) to the lumen of the stomach (pH 1.0) against a concentration gradient of greater than 10^6. This phenomenon has been a subject for experiment and speculation for decades, but a breakthrough was achieved in 1974 by Lee, Simpson and Scholes [83] with the following experiment.

Membrane vesicles were obtained from gastric mucosa by homogenization and were suspended in KCl at pH 6. The addition of ATP produced an alkalification of the medium. The proposed explanation was that the vesicles represent everted portions of the plasma membrane or portions of the 'tubulo–vesicular system' of the parietal cell not yet fused with the cell membrane (Fig. 6.1). ATP presented to the outside of the vesicle (corresponding to the cytoplasmic side of the cell membrane) caused H^+ ions to be pumped into the vesicle (i.e. corresponding to the outside of the cell). It was necessary first to soak the vesicles in KCl (or ^{86}Rb ^{36}Cl) from which it was deduced that the 'acid pump' of the parietal cell operates an ATP-driven exchange of H^+ for K^+. This was confirmed by measuring the ATP-dependent outflow of $^{86}Rb^+$ which was commensurate with the H^+ inflow. The ^{36}Cl did not move, indicating an electroneutral pump.

Hormonal stimulation of acid secretion by histamine causes an increase in cAMP, a disappearance of membrane from the 'tubulo–vesicular system' and a simultaneous increase in the surface area of the microvilli. It looks as though new pump molecules are being recruited to the cell membrane from a store of ready-formed membranes in the cell, (c.f. Section 6.2).

This H^+/K^+ exchange pump occurs only in the acid-secreting regions of the gastric mucosa. It appears around the 23rd day in the development of

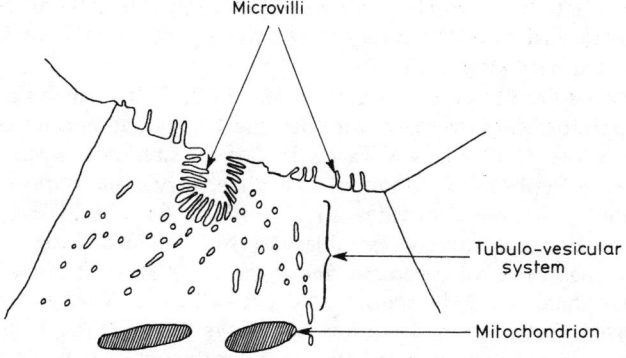

Fig. 6.1 Diagramatic section of a parietal (oxyntic) cell of the stomach.

rabbit embryo and at the same time as acid secretion is first observed. Fluorescent anti-enzyme antibodies localize the enzyme on the linings of the secretory canaliculus of the parietal cell. So there is little doubt about its role in acid secretion.

There are two or three puzzles though. How does the K^+ that is required for the H^+/K^+ exchange get into the lumen? Why does acid secretion in intact tissue require Cl^- (or other permanent anions such as Br^- or NO_3^-) when the vesicle preparation of Lee, Simpson and Scholes does not? The acid glands secrete more than 20 ml of fluid per hour of isotonic HCl, whereas the $(H^+ + K^+)$-ATPase alone probably causes no volume flow. It is tempting to suggest that all these points may be related. An outward movement of K^+ through a specific porter down its concentration gradient could be accompanied passively by Cl^- diffusion. This outward movement of salt could drive the volume flow, by causing the passive osmotic movement of water. The $(H^+ + K^+)$ATPase could then exchange K^+ for H^+, converting the bulk of the KCl to HCl (Fig. 6.2).

Purified preparations of an enzyme showing K^+-stimulated ATPase activity have been obtained from gastric mucosal tissue. In several respects this acid-pump shows a considerable resemblance to the $(Na^+ + K^+)$-ATPase. Electrophoresis in dissociating conditions indicates more than 90% of the protein in a single band of M_r 100 000. However, it appears that there are three types of protein, which must therefore all be of around M_r 100 000. Two thirds of the protein is degraded by trypsin, but a M_r 100 000 glyco-

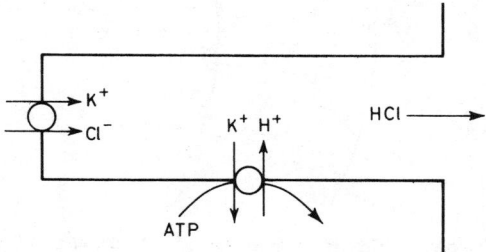

Fig. 6.2

protein is left intact. In the presence of ATP only one third of the protein is degraded and both the catalytic and the glycoprotein subunits survive trypsin treatment (Fig. 6.3).

ATP phosphorylates one residue per M_r of 625 000, though there appear to be approximately two nucleotide binding sites in that amount of protein. (As with the $(Na^+ + K^+)$-ATPase, 'half-of-the-sites' reactivity has been suggested.) Dephosphorylation of the phosphorylated protein is greatly stimulated by monovalent cations $(K^+ > Rb^+ > Cs^+, NH_4^+)$, so in this way also the acid pump strongly resembles the $(Na^+ + K^+)$-ATPase.

There has been disagreement about the H^+/ATP stoichiometry. Sachs says that there are 2 H^+ translocated per *extra* (i.e. K^+-stimulated) ATP hydrolysed. Scholes, on the other hand says that it is the $H^+/total$ ATP hydrolysed that is relevant and that it is very close to 1, while the H^+ to *extra* ATP is high and very variable, and that it is dependent on the K^+ concentration. Two further arguments favour the 1:1 stoichiometry. Scholes and colleagues showed [84] that the ATP hydrolysis in the absence of K^+ was nevertheless due to the acid pump: the result of a spontaneous breakdown of E—P to E in the absence of alkali cation. The second argument is a thermodynamic one. If the ΔG of ATP hydrolysis in the conditions of the oxyntic cell cytoplasm is -50 kJ mol^{-1}, a maximum ionic gradient of $10^{8.7}$ is possible. ($\Delta G = 2.3\ RT \log C_1/C_2$). Remembering that K^+ is pumped (electroneutrally) into the cell against a 10-fold gradient, that leaves $10^{7.7}$ available for H^+ pumping. If two H^+ were pumped per ATP they could only be pumped against a $10^{3.9}$ fold gradient, whereas H^+ gradients of greater than 10^6 are found. A stoichiometry of 1 H^+:1 K^+:1 ATP seems likely.

The $(H^+ + K^+)$-ATPase is inhibited by vanadate and by sulphydryl-reactive agents (e.g. Hg^{2+}), as is the $(Na^+ + K^+)$-ATPase. However, it is not inhibited by SCN^- or oubain, which confirms that this pump is distinct both from the F_0F_1, H^+-ATPase of the mitochondrion (SCN-sensitive) and the familiar $(Na^+ + K^+)$-ATPase of the cell membrane (ouabain-sensitive).

Brief mention was made above of the control of acid secretion. It will

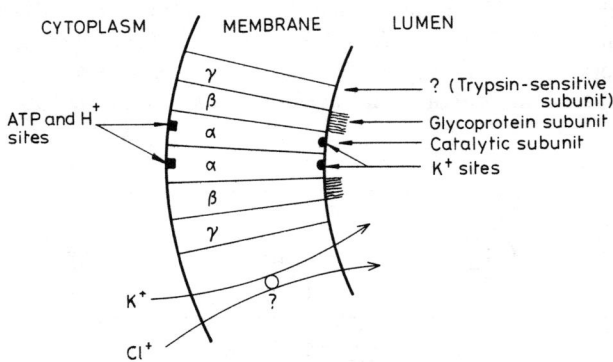

Fig. 6.3 Tentative hexameric structure of the $(H^+ + K^+)$-ATPase of gastric mucosa.

be clear to the reader that there is considerable medical and pharmaceutical interest in understanding this topic, because the excessive secretion of gastric acid and gastric ulceration are increasingly common stress symptoms in our modern stressful way of life.

6.2 Diabetes mellitus and the normal function of insulin

The disease we call diabetes was known to the ancients but the variety *Diabetes mellitus* was distinguished and named two centuries ago by an observant physician who noticed that the urine of sufferers tasted sweet. It is now well known that the immediate lesion is in most cases a failure of the pancreas to produce or release sufficient insulin into the blood stream, and that the fundamental consequence of insufficient insulin production by the pancreas is a failure of the cells of certain tissues to take up extracellular glucose. Insulin, of course, also regulates several intracellular enzymes concerned with the regulation of fat and carbohydrate metabolism, topics that are outside the scope of this monograph; it is only the effect of insulin on the transport of glucose into the cells of muscle and adipose-tissue that will be considered here.

Within minutes of adding insulin to the medium in which isolated adipocytes are incubated it is possible to demonstrate an increased uptake of the non-metabolizable glucose analogue 3-O-methylglucose ($t_{1/2}$ = 4 min.). The question then is — has the affinity of the carrier for glucose changed, has the turnover number changed or has there, in that short space of time, been an increase in the number of glucose carrier molecules in the plasma membrane?

That problem was finally resolved after the development of sensitive assays for the glucose carrier. Cytochalasin B (Fig. 3.10) is known to bind to several cellular constituents and to be a potent inhibitor of glucose transport (cf. Section 3.2.2). Low concentrations of radioactive cytochalasin B therefore bind to the glucose carrier, though also to other sites. Cushman and Wardzala [85] rather bravely suggested that the glucose-inhibitable fraction of cytochalasin B binding (400 mM glucose, 1 µM cytochalasin B) represented that bound to glucose carriers. They demonstrated a very close relationship between the rate of 3-O-methyl-glucose transport and the number of glucose carriers present in the plasma membrane fraction using this assay.

Insulin, at 0.7 nM causes a 4-fold increase in the number of glucose carriers in the plasma membrane even more rapidly ($t_{1/2}$ = 2.5 min.) than the observed increase in transport activity. With isolated adipocytes, very low concentrations of insulin will cause these effects (50% response to 0.1 nM insulin). However, with intact diaphragm muscle the same phenomena can be demonstrated with a 1000-fold higher insulation concentration [86].

Suzuki and Kono [87] succeeded in assaying the glucose carrier by reconstituting it into phospholipid vesicles and observed the same responses to insulin, i.e. an increase in the plasmamembrane. If insulin is removed, for example by adding excess anti-insulin antibody, the transport and the number of carriers in the cell membrane both decline with a $t_{1/2}$ of 9 min. These rapid responses to insulin were explained when it was found by Suzuki and Kono [87] that there is a pool of membrane in the cell in the form of 'golgi-like' vesicles that contains glucose carrier. It seems that insulin, or some intracellular consequence of insulin-binding to the plasma membrane, causes the fusion of that intracellular membrane with the plasma membrane in a reversible process akin to exocytosis/endocytosis. The transfer of individual carriers out of one bilayer and into the other has not been ruled out, though it seems less likely. Furthermore, the above-mentioned observations on the oxyntic cell (Section 6.1) suggest the feasibility of the former mechanism. Indeed, it is very interesting to see the strong similarity emerging between the way in which insulin increases glucose inflow into the glucose-storing tissues, and the way in which hormones increase acid secretion in the stomach.

6.3 Cholera

It has been known for many years that the fatal effects of cholera toxin were a result of excessive water loss from the intestine. The treatment of cholera is simply to keep the patient hydrated by intravenous infusion of saline, or oral administration of saline to which glucose has been added to aid solute, and with it, solvent, uptake. Most biological membranes have a high water permeability and it seems to be a general rule that water is not pumped; solutes are pumped and water follows passively by osmosis. Water loss is thus a passive consequence of salt or other solute loss. The question thus becomes: 'How does cholera toxin allow (or cause) solute loss?'. And the answer is still unsatisfactory.

Cholera toxin, a protein of M_r 87 000, has recently been shown to contain two types of subunit, A and B. The B peptides, of which there are 5 copies per molecule, bind to the hydrophilic headgroups of GM_1 gangliosides on the surface of susceptible cells, whereupon the A subunit is injected into the cell. There it leads to a large rise in cAMP concentration, by covalently modifying the 'G-protein', i.e. the GTP-binding and hydrolysing regulatory subunit of adenyl cyclase. The mechanism of this covalent modification is a bit bizarre though not unique to cholera toxin. The A peptide is cleaved by disulphide reduction in the cell revealing a NAD-glycohydrolase enzymic activity that transfers ADP ribose from NAD to the 'G-protein'. The ADP-ribosylated G-protein is thus locked in its GTP-binding conformation in which it activates the adenyl cyclase. This aspect of cholera action has been reviewed by van Heynigen [88] and is summarized in Fig. 6.4.

Between the stomach and the anus it seems that Na^+ and Cl^- are first pumped out into the lumen of the gut, by the upper small intestine, and then transported back out of the lumen in the colon, in order to concentrate the faeces. Raised cAMP certainly stimulates secretion in other secretory tissues such as the exocrine pancreas, salivary gland, sweat gland, etc., and though there is little firm evidence it seems likely that the diarrhoea of cholera is due to increased NaCl pumping in the jejunum rather than impaired NaCl transport in the colon.

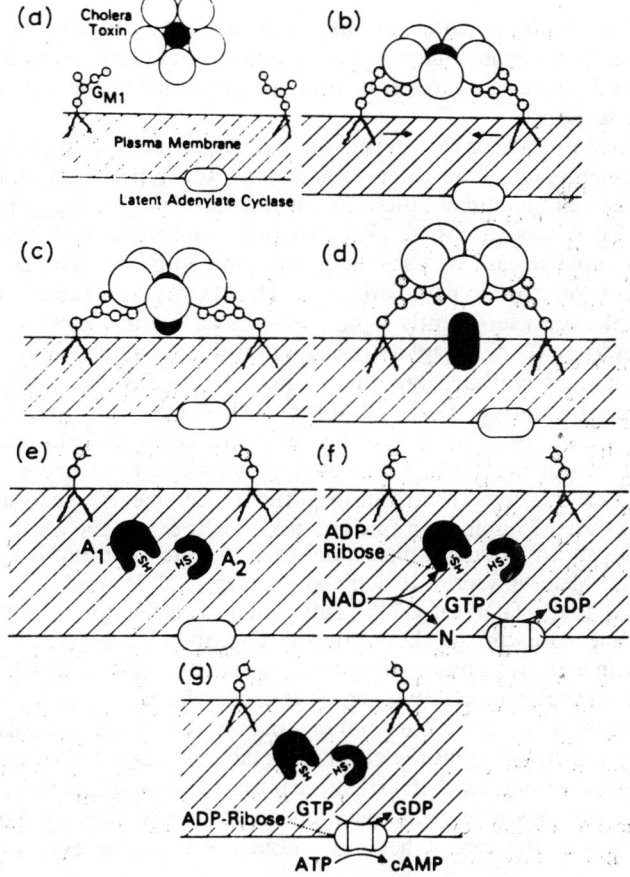

Fig. 6.4 Cholera toxin stimulates adenyl cyclase (from [89] with permission).

There is still no information on the ion pumps or channels involved or the link between cAMP and these ion-porters. It seems that the motive-power for jejunal secretion is the oubain-sensitive $(Na^+ + K^+)$-ATPase, but that does not solve the problem, for it seems that that enzyme is confined to the serosal side of the cell and therefore does not vent its Na^+ out into the lumen of the gut but into the plasma. This aspect of the action of cholera toxin has been reviewed by Field [89].

6.4 Vision

Bacteriorhodopsin (Chapter 5) was so called because it bears a strong resemblance to the well-known visual pigment rhodopsin; that is to say, both pigments bear the identical chromophore in identical Schiff-base linkage to an intrinsic membrane protein. As bacteriorhodopsin turned out to be a light-driven proton pump, the question that springs to mind is whether rhodopsin acts in the eye in a similar way, i.e. itself producing a light-driven ion current that hyperpolarizes the rod-cell-membrane, thus causing an action-potential in the nerve. According to current thinking rhodopsin is not an ion pump, but a brief consideration of vision is justified as it reveals some very interesting overlaps with some of the topics covered in other sections.

The first problem is that in the rod cell, rhodopsin is sited exclusively in the disc membrane of the outer-segment while the hyperpolarization occurs across the cell membrane; these two membranes are topologically distinct (Fig. 6.5). It was suggested [90] that rhodopsin when illuminated could cause a rapid release of Ca^{2+} from the discs into the cytoplasm which could in turn cause hyperpolarization. That theory also turned out to be too simple, for a sufficiently rapid release of Ca^{2+} into the cytosol has not been observed. (Ca^{2+} *is* rapidly released into the disc lumen [94].)

One current working hypothesis, that of Bitensky and others, is well summarized by Stryer [91] and Miller [92]. There seems little doubt that this mechanism operates at very low light intensities in dark-adapted rods and that it can explain how one photon can block the entry of 10^6 Na^+ ions and cause a hyperpolarization of the rod-cell-membrane in the period 0.1–2.0 s after illumination (see Fig. 6.5). Light causes the transformation of rhodopsin to metarhodopsin I and then to metarhodopsin II with a $t_{1/2}$ of 5 ms. Metarhodopsin II somehow causes a GTP-binding protein to exchange bound GDP for GTP. (This protein is called the G-unit or transducin; but note the parallel with the G-protein in Section 6.3.) GTP-transducin activates a phosphodiesterase specific for cGMP, which proceeds to lower cytoplasmic cGMP levels. That causes the closing of Na^+-conductance channels in the cell membrane and results in a hyperpolarization. (Injection of cGMP opens the channels again and causes depolarization.)

It might be argued that this mechanism, though exquisitely sensitive and relevant to low-level illumination might be two slow for bright-light vision; and in any case it has not considered the early events (0–1 μs)

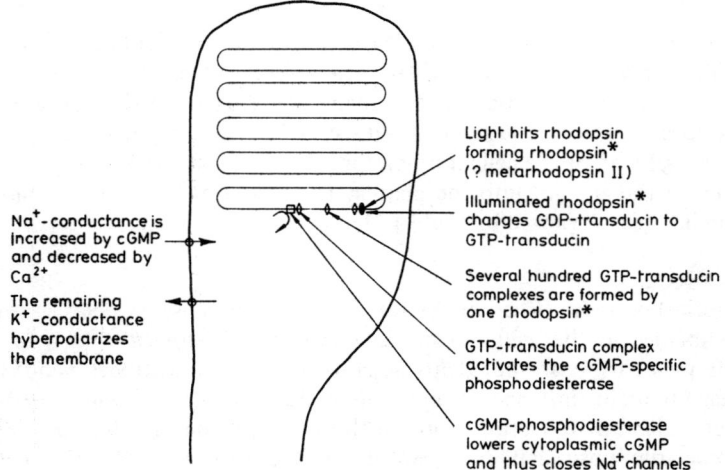

Fig. 6.5 One working hypothesis for the visual process in the rod outer segment.

linking the absorption of the photon with the conformational changes of the rhodopsin molecule (Fig. 6.5).

If electrodes are placed on the optic nerve or across isolated whole retinas and a very short (5 ns) laser flash given to the retinal surface, electrical signals are observed in the nerve in the time-domain 0.1 μs to 1 ms. These are called 'early receptor photovoltages'. There are two suggestions as to their origin. (1) Rhodopsin might after all be a proton pump like bacteriorhodopsin and the release of a proton might cause the photovoltage, followed by a protein conformational change as a result of deprotonation. (2) A *cis* → *trans* isomerization in the illuminated retinal chromophore might force apart a salt-bridge in the centre of the molecule, and that physical movement of charge might cause the small electrical signals [93].

However, Ca^{2+} does enter the story again in two ways.

(1) A flash of light does cause the release of Ca^{2+} from the cell in sufficient quantity and with sufficient speed to qualify it for a role in hyperpolarization. One suggestion is that Ca^{2+} acts between cGMP and the Na^+-channel.
(2) Kaupp et al. [94] report that with the same time course as the transition from metarhodopsin I to metarhodopsin II, there is the uptake of 2.8 H^+ and the release of 1 Ca^{2+} into the disc lumen, (c.f. bacteriorhodopsin). There is, however, no obvious link as yet between this process and hyperpolarization.

6.5 Control of ion-flow through nerve membranes

Two further, and rather marvellous, examples of the control of transport are afforded by studies on the mechanism of nervous transmission. A good beginning has been made towards understanding the biochemistry of both the sodium channel and the acetylcholine regulator (i.e. receptor and ion channel).

6.5.1 The Na^+ channel

The nerve impulse consists of two processes: a depolarization caused by the flow of Na^+ into the nerve, followed by a repolarization caused by the flow of K^+ out of the nerve and, at the same time, the closing of the Na^+-channels. However, that initial opening of the Na^+-channel is caused by a depolarization, the very thing that it is itself the cause of, and in this way the nerve-impulse is self-propagating. A certain amount is now known of the biochemistry of the Na^+-channel; much less of the K^+-channel.

The Na^+-channel diameter can be gauged from its conductivity towards a series of cations of known ionic radius [97]. It appears that the channel presents an opening of dimensions 0.3 x 0.5 nm. Hydrated Li^+ and Na^+ penetrate readily. Of the following similarly sized amines $OHNH_3^+$ and $NHNH_2NH_3^+$ pass, but $CH_3NH_3^+$ is excluded, so size is clearly not the only consideration. From this it seems that the electronegative and hydrogen-bond-forming atom at the left of the first two ions is involved, perhaps in hydrogen-bond formation. K^+ passes 12 times slower than Na^+. But K^+ is only larger than $NH_2NH_3^+$ if the K^+ is partially hydrated, (Fig. 6.6). Both these arguments indicate that the hydrated Na^+ is the substrate.

The approximate molecular weight of the channel protein has been estimated by the application of target theory to data on the inactivation of the Na^+-channel by high-energy electrons [98]. The basis of this technique is that it is supposed that the absorption of the energy of a single electron will inactivate the entire protein molecule that is hit but will cause no dammage to neighbouring molecules. Though the technique is well vindicated for soluble proteins in aqueous solution, the technique is often applied to membrane proteins *in situ*, when no other means of measuring the molecular dimensions exists, so its validity should not be too heavily relied upon. The Na^+-channel appears, by this technique, to be of M_r 230 000.

There is a very specific and tightly-binding inhibitor called tetrodotoxin, that is produced by several rather unrelated animals but most notably by the Japanese puffer fish. It is, of course, intensely poisonous, causing paralysis, and there are several deaths per year that result from the careless preparation and cooking of this otherwise excellent and highly esteemed fish. The tightness and the specificity of tetrodotoxin binding has justified the isolation of a tetrodotoxin-binding protein from the electroplax of the electric eel that is assumed to be the Na^+-channel protein. This protein turns out to have a molecular weight similar to that obtained by electron inactivation.

One of the most interesting features of the Na^+-channel is of course its control by the membrane potential. There must presumably be a conformational change that is a consequence of the movement of positive or negative charges in the plane of the field. The biochemistry of this is still completely obscure, but the movement of three or four charges per tetrodotoxin binding site has been detected electrophysiologically following a step change in applied membrane potential [98].

6.5.2 The acetylcholine regulator

The acetylcholine-activated ion-channel of the post-synaptic membrane of certain types of neuromuscular junction has been studied and reviewed by Changeux and colleagues [99]. They use the term acetylcholine regulator to describe the whole complex, i.e. the acetylcholine receptor protein and the ion-channel that it regulates (Fig. 6.7).

The regulator complex can be isolated in mg amounts from the electric

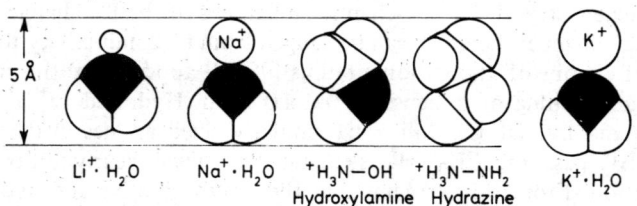

Fig. 6.6 The ionic selectivity of the sodium channel partly depends on steric factors. Na^+ and Li^+, together with a water molecule, fit in the channel, as do hydroxylamine and hydrazine. In contrast, K^+ with a water molecule is too large (from [98] with permission from Scientific American, Inc).

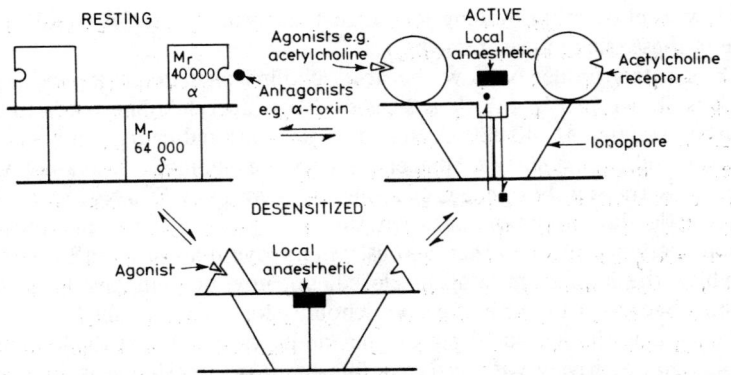

Fig. 6.7 Figurative representation of the acetylcholine regulator of the fish electroplax and the motor end-plate of the neuromuscular junction (from [99]).

organs of torpedo rays and electric eels, where it constitutes up to 40% of the membrane protein, as well as from skeletal muscle. The biochemical study of the regulator complex, as with the Na^+-channel, was made possible by the discovery of very tightly binding toxins. In this case the toxins are the α-toxins of certain poisonous snakes such as *Naja sp* and *Bungarus sp.*, which are small peptides of 60—70 amino acids. These are found to bind to the acetylcholine receptor with a dissociation constant of 10^{-11} M (implying a very slow rate of dissociation). Toxins such as α-bungarotoxin have been used both to monitor the progress of standard purification procedures, such as differential centrifugal fractionation of electric organ membranes, and also as an affinity ligand for affinity separation procedures. The usual non-ionic detergents (Triton X-100, Tween 80, etc.) are necessary for the removal of regulator complex from the membrane and its separation into receptor and ionophore. The purest preparations of complex bind 4—5 μmol of α-bungarotoxin per g protein, corresponding to M_r of 200 000 —250 000 per toxin binding site. With isolated receptor subunit a figure of 100 000 daltons per binding-site is obtained.

SDS-polyacrylamide gel electrophoresis of purified complex shows four bands: 40 000, 50 000, 57 000 and 64 000 M_r referred to as the $\alpha\beta\gamma\delta$ subunits respectively [100]. If the subunits are first separated with detergents, the toxin-binding protein is found to be the 40 000 M_r peptide. It is suggested that the 64 000 M_r peptide might be the site at which local anaesthetics bind, but there is as yet little indication as to the function of the other subunits, and which subunit is the ionophore. Electron micrographs and sedimentation studies suggest a possible pentameric or hexameric structure e.g. $\alpha_2\beta\gamma\delta$ (shown figuratively as a $\alpha_2\delta_2$ tetramer in Fig. 6.7 as there is no information on the role of the β and γ subunits). It seems likely therefore that each subunit M_r 40 000, contains a single site that can bind the α-toxin but with such a low 'off' rate that binding can hardly be said to be reversible. The same site can reversibly, and thus competitively, bind acetylcholine, or analogous quaternary amines, and also d-tubocurarine, one active ingredient of the amazonian indian's arrow poison. (Half-of-the-sites reactivity has once again been talked of to explain the low level of toxin binding to purified receptor, but errors could easily arise in these tricky measurements.)

A tentative model of how the acetylcholine regulator complex might work is shown in Fig. 6.7. It is suggested that acetylcholine binds to the receptor subunit M_r 40 000 causing a conformational change in both the receptor subunit and the adjacent ionophore subunit. The ionophore channel is transiently opened, allowing the passage of K^+, Na^+ and Ca^{2+} (with little discrimination between the first two). But the presence of bound acetylcholine (or other 'agonist') must favour a third conformation, in which the ionophore is again closed, but now not sensitive to acetylcholine, because, of course, the acetylcholine site is already filled.

Local anaesthetics, such as procaine and lignocaine, block the ion-channel non-competitively with acetylcholine and it is suggested that they bind at a different site, possibly on the ionophore itself (Fig. 6.7).

Conductance measurements have shown that the opening of the ion channel occurs within 1 ms and that it closes again with an exponential half-life of several ms. These measurements indicate the rate constants of the proposed conformational changes [101].

6.6 Hereditary defects of transport

As with any other sort of protein, mutation can produce defective transport carriers. A number of quite well recognized diseases are a result of such mutations; they are rare because in most cases these defects only show clinically in the homozygote. A few of the better characterized hereditary defects of transport will be listed. (For reviews see [95,96].)

(1) Hereditary chloride diarrhoea. This is a result of the loss of the Cl^-/HCO_3^- anion exchanger of the large intestine. The unusually high Cl^- content of the faeces prevents their normal dehydration.

(2) Hartnup disease. This condition is due to a loss of the neutral amino acid carrier of the small intestine. The results are not severe, though the uptake of a number of amino acids is affected. The patients survive quite well by absorbing oligopeptides. Lack of tryptophane causes nicotinamide deficiency, but it can be administered orally.

(3) Cystinuria. This is caused by loss of the carrier for Lys, Arg and Cys from both intestinal and renal brush borders. The ill effects are largely a result of kidney stones formed from the sparingly soluble cystine.

(4) Glucose-galactose malabsorption. Glucose and galactose apparently share a common Na^+-dependent, coupled active transport system which is present in both intestine and kidney. However, glucosuria is slight, and it is clear that there is a second carrier specific to the kidney. Biopsy, via the mouth, of mucosa from the jejunum has shown an absence of glucose and galactose accumulation, compared with the 16-fold accumulation in normal mucosa. Crane [95] has further shown a 90% decrease in the Na^+-dependent phlorizin binding, from 91–8 nm (ml tissue)$^{-1}$.

(5) Fanconi syndrome, in which all Na^+-coupled symporters of the brush borders lose their power to accumulate substrates.

Appendix

Following the King–Altman procedure, one can quite simply write down the rate equation, as follows. (See Engel [17] for further detail. You need not understand the King–Altman procedure in order to apply it.) First write down a statement which defines the relative abundance of each of the four carrier species in the steady state:

$$\frac{C_o}{\rightarrow|+\uparrow|+\downarrow|+\swarrow} = \frac{CS_o}{\rightarrow|+\uparrow|+\swarrow|+\downarrow|} = \frac{CS_i}{\swarrow|+\uparrow|+\swarrow|+\uparrow|}$$

$$= \frac{C_i}{\downarrow|+\swarrow|+\swarrow|+\uparrow|+\rightarrow|} = \frac{C_o}{D_1} = \frac{CS_o}{D_2} = \frac{CS_i}{D_3} = \frac{C_i}{D_4}$$

(if D_1, D_2, etc., stand for those denominators). Then write down the conservation equation:

$$C_T = C_o + CS_o + CS_i + C_i$$

(where C_T is the total carrier concentration).

We can now express C_T in terms of any one of the carrier species, e.g. C_i, C_o, etc., as follows

$$C_T = \frac{C_i(D_1 + D_2 + D_3 + D_4)}{D_4} = \frac{C_o(D_1 + D_2 + D_3 + D_4)}{D_1} \text{ etc.}$$

or $$C_i = C_T \frac{D_4}{(D_1 + D_2 + D_3 + D_4)}$$

$$C_o = C_T \frac{D_1}{(D_1 + D_2 + D_3 + D_4)}$$

etc.

But the velocity (v) of net inflow of substrate is $C_i g - C_o h$, just as it is also $CS_o c - CS_i d$, etc.

Therefore,

$$v = \frac{C_T D_4 g}{(D_1 + D_2 + D_3 + D_4)} - \frac{C_T D_1 h}{(D_1 + D_2 + D_3 + D_4)} = C_T \frac{(D_4 g - D_1 h)}{(D_1 + D_2 + D_3 + D_4)}$$

At last we have to write out the D terms in full in order to simplify and cancel.

So,

$$\frac{v}{C_T} = \frac{gehb + gech + geca\,[S_o] + gdbh - hdbg - hegb - hceg - hfdb\,[S_i]}{\begin{array}{l}bdg + bdf\,[S_i] + beg + ceg + dga\,[S_o] + daf\,[S_o][S_i] + hdf\,[S_i]\\ + aeg\,[S_o] + cfh\,[S_i] + bdh + bhf\,[S_i] + acf\,[S_o][S_i] + aec\,[S_o]\\ + acg\,[S_o] + ceh + beh\end{array}} \qquad (1)$$

Cancelling and collecting terms:

$$\frac{v}{C_T} = \frac{aceg\,[S_o] - bdfh\,[S_i]}{\begin{array}{l}(h+g)(bd+be+ce) + [S_i]f(bd+bh+ch+dh)\\ + [S_o]a(ce+cg+dg+eg) + [S_i][S_o]\,af(c+d)\end{array}} \qquad (2)$$

For the initial rate of influx with no substrates on the inside (often called 'zero-trans' influx) we can omit terms in $[S_i]$:

therefore

$$v = \frac{C_T[S_o]aceg/\{a(ce+cg+dg+eg)\}}{[S_o] + (h+g)(bd+be+ce)/\{a(ce+cg+dg+eg)\}} \qquad (3)$$

This is Equation (2.5) given on page 25.

References

[1] Overton, E. (1899), Vierteljahrsschr. Naturforsch. Ges. Zürich. *Braunschweig.*, **44**, 88–107.
[2] Quinn, P. J. (1976), *The Molecular Biology of Cell Membranes*, MacMillan Press, London.
[3] Tanford, C. (1980), *The Hydrophobic Effect: Formation of Micelles and Biological Membranes*, John Wiley and Sons, New York.
[4] Oparin, A. I. (1962), *Life: Its Nature, Origin and Development*, Academic Press, New York.
[5] Ling, G. N. (1962), *A Physical Theory of the Living State*, Ginn, Boston.
[6] Troshin, A. S. (1966), *Problems in Cell Permeability*, Pergamon Press, Oxford.
[7] Kleinzeller, A. and Kotyk, A. (1961), *Membrane Transport and Metabolism*, Academic Press, New York.
[8] Sistrom, W. R. (1958), *Biochim. Biophys. Acta*, **29**, 579–587.
[9] Mitchell, P. (1970), *Symp. Soc. Gen. Microbiol.*, **20**, 121–166.
[10] Heller, K. B., Lin, E. C. C. and Wilson, T. H. (1980), *J. Bacteriol.*, **144**, 274–278.
[11] Sharp, A. P. and Thomas, R. C. (1981), *J. Physiol.*, **312**, 71–80.
[12] Stein, W. D. (1967), *The Movement of Molecules Across Cell Membranes*, Academic Press, London.
[13] Mitchell, P. (1967), *Adv. Enzymol.*, **29**, 33–87.
[14] Lieb, W. R. and Stein, W. D. (1971), *Nature New Biology*, **234**, 220–222.
[15] Lieb, W. R. and Stein, W. D. (1974), *Biochim. Biophys. Acta*, **373**, 165–177.
[16] Schultz, S. G. (1980), *Basic Principles of Membrane Transport*, Cambridge University Press, Cambridge.
[17] Engel, P. C. (1977), *Enzyme Kinetics* (Outline Studies in Biology), Chapman and Hall, London.
[18] Sen, A. K. and Widdas, W. F. (1962), *J. Physiol.*, **160**, 392–403.
[19] Lieb, W. R. and Stein, W. D. (1972), *Biochim. Biophys. Acta*, **265**, 187–207.
[20] Bloch, R. (1974), *J. Biol. Chem.*, **249**, 3543–3550.
[21] Regen, D. M. and Tarpley, H. L. (1974), *Biochim. Biophys. Acta*, **339**, 218–233.
[22] Page, M. G. P. and West, I. C. (1982), *Biochem. J.*, **204**, 681–688.
[23] Mitchell, P. (1977), *Symp. Soc. Gen. Microbiol.*, **27**, 383–423.
[24] DiRienzo, J. M., Nakamura, K. and Inouye, M. (1978), *Ann. Rev. Biochem.*, **47**, 481–532.
[25] Lugtenberg, B. (1981), *Trends. Biochem. Sci.*, **6**, 262–266.
[26] Nakae, T. (1976), *Biochem. Biophys. Res. Commun.*, **71**, 877–884.

[27] Chen, R., Krämer, C., Schmidmayr, W. and Henning, U. (1979), *Proc. Natl. Acad. Sci.* (USA), **76**, 5014–5017.
[28] Benz, R., Janko, K. and Läuger, P. (1979), *Biochim. Biophys. Acta*, **551**, 238–247.
[29] Ferenci, T., Brass, J., Boos, W. (1980), *Biochem. Soc. Trans.*, **8**, 680–681.
[30] Knauf, P. A. (1979), *Curr. Top. Membr. Transp.*, **12**, 249–363.
[31] Cabantichik, Z. I., Knauf, P. A. and Rothstein, A. (1978), *Biochim. Biophys. Acta*, **515**, 239–302.
[32] Wieth, J. O. (1979), *J. Physiol.*, **294**, 521–539.
[33] Rothstein, A., Ramjeesingh, M., Grinstein, S. and Knauf, P. A. (1980), *Ann. N.Y. Acad. Sci.*, **341**, 433–443.
[34] Williams, D. G., Jenkins, R. E. and Tanner, M. J. A. (1979), *Biochem. J.*, **181**, 477–493.
[35] Batt, E. R., Abbot, R. E. and Schachter, D. (1976), *J. Biol. Chem.*, **251**, 7184–7190.
[36] Kasahara, M. and Hinkle, P. C. (1976), *Proc. Natl. Acad. Sci.* (USA), **73**, 396–400.
[37] Baldwin, S. A. and Lienhard, G. E. (1981), *Trends Biochem. Sci.*, **6**, 208–211.
[38] Kasahara, M. and Hinkle, P. C. (1977), *J. Biol. Chem.*, **252**, 7384–7390.
[39] Nickson, J. K. and Jones, M. N. (1982), *Biochim. Biophys. Acta*, **690**, 31–40.
[40] Gorga, F. R. and Lienhard, G. E. (1982), *Biochemistry*, **21**, 1905–1908.
[41] Crane, R. K. (1977), *Rev. Physiol. Biochem. Pharmacol.*, **78**, 99–159.
[42] West, I. C. (1980), *Biochim. Biophys. Acta*, **604**, 91–126.
[43] Fox, C. F. and Kennedy, E. P. (1965), *Proc. Natl. Acad. Sci.* (USA), **54**, 891–899.
[44] West, I. C. (1970), *Biochem. Biophys. Res. Commum.*, **41**, 655–661.
[45] West, I. C. and Mitchell, P. (1972), *J. Bioenerg.*, **3**, 445–462.
[46] West, I. C. and Mitchell, P. (1973), *Biochem. J.*, **132**, 587–592.
[47] Teather, R. M., Müller-Hill, B., Abrutsch, U., Aichele, G. and Overath, P. (1978), *Mol. Gen. Genet.*, **159**, 239–248.
[48] Rickenberg, H. V., Cohen, G. N., Buttin, G. and Monod, J. (1956), *Ann. Inst. Pasteur.*, **91**, 829–857.
[49] Mitchell, P. (1963), *Biochem. Soc. Symp.*, **22**, 142–169.
[50] Pavlasova, E. and Harold, F. M. (1969), *J. Bacteriol.*, **98**, 198–204.
[51] West, I. C. (1973), in *Ion Transport in Plants* (ed. W. P. Anderson), Academic Press, London, pp. 237–250.
[52] Booth, I. R., Mitchell, W. J. and Hamilton, W. A. (1979), *Biochem. J.*, **182**, 687–696.
[53] Zilberstein, D., Schuldiner, S. and Padan, E. (1979), *Biochemistry*, **18**, 669–673.
[54] Jones, T. H. D. and Kennedy, E. P. (1969), *J. Biol. Chem.*, **244**, 5981–5987.
[55] Büchel, D. E., Gronenborn, B. and Müller-Hill, B. (1980), *Nature*, **283**, 541–545.
[56] Ehring, R., Beyreuther, K., Wright, J. K., and Overath, P. (1980), *Nature*, **283**, 537–540.

[57] Beyreuther, K., Bieseler, B., Ehring, R., Griesser, H., Mieschendahl, M. and Müller-Hill, B. (1980), *Biochem. Soc. Trans.*, **8**, 675–676.
[58] Beyreuther, K., Bieseler, B., Ehring, R. and Müller-Hill, B. (1981), in *Methods in Protein Sequence Analysis* (ed. M. L. Elsinga), Humana Press, Clifton, N. J.
[59] Eddy, A. A. (1980), *Biochem. Soc. Trans.*, **8**, 271–273.
[60] Booth, I. R. (1981), *Trends Biochem. Sci.*, **6**, 257–262.
[61] Page, M. G. P. and West, I. C. (1982), *Biochem. J.*, **204**, 681–688.
[62] Skou, J. C. (1957), *Biochim. Biophys. Acta*, **23**, 394–401.
[63] Jørgensen, P. L. (1982), *Biochim. Biophys. Acta*, **694**, 27–68.
[64] Esman, M. (1980), *Anal. Biochem.*, **108**, 83–85.
[65] Deguchi, N., Jørgensen, P. L. and Maunsbach, A. B. (1977), *J. Cell. Biol.*, **75**, 619–634.
[66] Freytag, J. W. and Reynolds, J. A. (1981), *Biochemistry*, **20**, 7211–7214.
[67] Skou, J. C. (1975), *Quart. Rev. Biophys.*, **7**, 401–434.
[68] Post, R. L, Hegyvary, C. and Kume, S. (1972), *J. Biol. Chem.*, **247**, 6530–6540.
[69] Albers, R. W. (1967), *Annu. Rev. Biochem.*, **36**, 727–756.
[70] Wyman, J. (1965), *J. Molec. Biol.*, **11**, 631–644.
[71] Fersht, A. (1977), *Enzyme Structure and Mechanism*, Freeman, Reading, U.K.
[72] Cantley, L. C., Cantley, L. G. and Josephson, L. (1978), *J. Biol. Chem.*, **253**, 7361–7368.
[73] Levit, D. G. (1980), *Biochim. Biophys. Acta*, **604**, 321–345.
[74] Karlish, S. J. D., Yates, D. W. and Glynn, I. M. (1978), *Biochim. Biophys. Acta*, **525**, 252–264.
[75] Karlish, S. J. D. and Yates, D. W. (1978), *Biochim. Biophys. Acta*, **527**, 115–130.
[76] Jørgensen, P. L., Karlish, S. J. D. (1980), *Biochim. Biophys. Acta*, **597**, 305–317.
[77] Davson, H. (1970), *A Textbook of General Physiology*, 4th edn, Churchill, London.
[78] Spector, M., O'Neal, S. and Racker, E. (1981), *J. Biol. Chem.*, **256**, 4219–4227.
[79] Henderson, R. and Unwin, P. N. T. (1975), *Nature*, **257**, 28–32.
[80] Khorana, H. G., Gerber, G. F., Herlihy, W. C., Gray, C. P., Anderegg, R. J., Nihel, K. and Biemann, K. (1979), *Proc. Natl. Acad. Sci. (USA)*, **76**, 5046–5050.
[81] Ovchinnikov, Y. A., Abdulaev, N. G., Feigina, M. Y., Kiselev, A. V. and Lobanov, N. A. (1979), *FEBS Lett.*, **100**, 219–224.
[82] Merz, H. and Zundel, G. (1981), *Biochem. Biophys. Res. Commun.*, 540–546.
[83] Lee, J., Simpson, G. and Scholes, P. (1974), *Biochem. Biophys. Res. Commun.*, **54**, 690–696.
[84] Smith, G. S. and Scholes, P. B. (1982), *Biochim. Biophys. Acta*, **688**, 803–807.
[85] Cushman, S. W. and Wardzala, L. J. (1980), *J. Biol. Chem.*, **255**, 4758–4762.
[86] Wardzala, L. J. and Jeanrenaud, B. (1981), *J. Biol. Chem.*, **256**, 7090–7093.

[87] Suzuki, K. and Kono, T. (1980), *Proc. Natl. Acad. Sci.* (USA), **77**, 2542–2545.
[88] Van Heyningen, S. (1981), *Bioscience Reports* 2, 135–145.
[89] Field, M. (1981), in *Physiology of the Gastrointestinal Tract.*, (ed. L. R. Johnson), Raven Press, New York, pp. 963–982.
[90] Hagins, W. A. and Yoshikami, S. (1975), *Ann. N. Y. Acad. Sci.*, **264**, 314–325.
[91] Stryer, L., Hurley, J. B. and Fung, B. K. K. (1981), *Trends Biochem. Sci.*, **6**, 245–247.
[92] Miller, W. H. (1981), *Current Topics in Membranes and Transport*, **15**, 441–445.
[93] Trissl, H. W. (1982), *Biophys. Struct. Mech.*, **8**, 213–230.
[94] Kaupp, U. B., Schretkamp, P. P. M. and Junge, W. (1981), *Biochemistry*, **20**, 5511–5516.
[95] Crane, R. K. (1980), *Biochem. Soc. Trans.*, **8**, 688–690.
[96] Stanbury, J. B., Wyngaarden, J. B., Fredrickson, D. S. and Goldstein, J. L. (eds) (1983), in *Metabolic Basis of Inherited Disease*, McGraw-Hill, New York, pp. 1729–1920.
[97] Hille, B. (1971), *J. Gen. Physiol.*, **58**, 599–619.
[98] Keynes, R. D. (1979), *Scientific American*, **240**, 126–135.
[99] Heidmann, T. and Changeux, J. P. (1978), *Ann. Rev. Biochem.*, **47**, 317–357.
[100] Gullick, N. J. and Lindstrom, J. M. (1982), *Biochemistry*, **21**, 4563–4568.
[101] Dionne, V. W. and Leibowitz, M. D. (1982), *Biophysical J.*, **39**, 253–261.
[102] Jones, C. W. (1981), *Biological Energy Conservation, Oxidative Phosphorylation*, Outline Studies in Biology, 2nd edn, Chapman and Hall, London.
[103] Engleman, D. M., Henderson, R., McLachlan, A. D. and Wallace, B. A. (1980), *Proc. Nat. Acad. Sci.* (USA), **77**, 2023–2027.
[104] Robbie, J. P. and Wilson, T. H. (1969), *Biochim. Biophys. Acta*, **173**, 234–244.
[105] Winkler, H. H. and Wilson, T. H. (1966), *J. Biol. Chem.*, **241**, 2200–2211.

Index

Acetylcholine regulator, 69, 70–72
Active transport, 10
Adipocytes, 65
ADP-ribosylation, 66, 67
Anion-exchange carrier, 29–35
Antiport, 22, 32
Na^+/H^+, 61
ATPase
$(Na^+ + K^+)$-ATPase, 47–57
$(H^+ + K^+)$-ATPase, 62–65

Bacteriorhodopsin, 57–61
Band 3 protein, 29–35
Band 4.5, glucose carrier, 36
Bilge pumps, 10

cAMP, 66
cGMP, 68, 69
Cholera, 66–67
Coomassie Blue stain, 30, 40, 48
Counter-transport, 18, 19
Cross-linker, see dimethylsuberimidate
Cu^{2+} oxidation of dithiols, 33, 50
Cystinuria, 72
Cytochalasin B, 35, 36, 65

Diabetes mellitus, 65
Diarrhoea
cholera, 66–67
hereditary chloride, 72
DIDS, 30, 31, 32
Dimethylsuberimidate, 28, 41, 50
Dinitrophenol, 38

Electric eel, 57, 70, 71
Electric field
kinetic effects, 24, 46
thermodynamic effects, 43–44
controls sodium channel, 70
Electrogenic pump, 55
Erythrocyte
anion-exchange carrier, 29–35
glucose transport, 35–37

N-ethylmaleimide
reactive cysteine of lactose carrier, 41
blocks $E_1 \to E_2$ conversion, 51
Exchange-diffusion, 18, 19

Facilitated diffusion
kinetics, 17–23
Fanconi syndrome, 72
Fick's Law, 14, 15

G-protein, 66
Gastric acid secretion, 62–65
Gated-pore, 17
Glucose carrier, 35–37
kinetic constants, 21, 22
rate constants, 37
Glucose-galactose malabsorption, 72
GM_1-ganglioside, 66, 67

Half-of-the-sites reactivity
$(Na^+ + K^+)$-ATPase, 54
$(H^+ + K^+)$-ATPase, 64
acetylcholine regulator, 71
Halorhodopsin, 61
Hamburger shift, 34
Hartnup's disease, 72
α-helix, 28, 41, 57, 59
Histamine stimulates secretion, 62
Hydrogen bonds, 16

Ionophore of acetylcholine regulator, 71
Insulin, 65, 66

Kinetics, 13–25
simple diffusion, 13–17
King-Altman procedure, 20, 23, 73

Lactose permease, 38
Lactose-proton symporter
defined physiologically, 38–39
DNA sequenced, 41, 42
Lam B protein, 29

79